# Regenerating Earth

**Farmers Working with Nature to Feed Our Future**

Kelsey Timmerman

patagonia®

## Regenerating Earth
Farmers Working with Nature to Feed Our Future

Patagonia publishes a select list of titles on wilderness, wildlife, and outdoor sports that inspire and restore a connection to the natural world and encourage action to combat climate chaos.

© 2025 Kelsey Timmerman

Photograph/Illustrator copyrights held by the photographer/illustrator as indicated in captions. Permission required for reproduction.

All rights reserved. No part of this book may be used or reproduced in any manner whatsoever without written permission from the publisher and copyright holders. Requests should be e-mailed to books@patagonia.com or mailed to Patagonia Books, Patagonia Inc., 259 W. Santa Clara St., Ventura, CA 93001-2717.

**Hardcover Edition**

**Published by Patagonia Works**

Printed in Canada on 100 percent postconsumer recycled paper.

Editor: Sharon AvRutick
Photo Editor: Jane Sievert
Art Director/Designer:
   Christina Speed
Project Manager: Sonia Moore
Photo Production:
   Bernardo Salce
Graphic Production:
   Michaela Purcilly
   Natausha Greenblott
Creative Director: Michael Leon
Publisher: Karla Olson

Hardcover ISBN
   9781952338267
E-Book ISBN
   9781952338274
Library of Congress Control Number 2025934863

COVER IMAGE: Farming successfully using regenerative practices in the extreme climate of Patagonia, Chile. CLIFF RITCHEY

FRONT ENDPAPER: "Why are churches sacred and rivers not?" Miguel, a member of the Arhuaco in Colombia, asked. CLIFF RITCHEY

TITLE PAGE: Volunteers harvest veggies at Faithfull Farms near Chapel Hill, North Carolina. FAITHFULL FARMS

### ENVIRONMENTAL BENEFITS STATEMENT
Patagonia Inc saved the following resources by printing the pages of this book on chlorine free paper made with 100% post-consumer waste.

| TREES | WATER | ENERGY | SOLID WASTE | GREENHOUSE GASES |
|---|---|---|---|---|
| 253 FULLY GROWN | 20,000 GALLONS | 106 MILLION BTUs | 800 POUNDS | 109,500 POUNDS |

 Environmental impact estimates were made using the Environmental Paper Network Paper Calculator 4.0. For more information visit www.papercalculator.org

FSC
www.fsc.org
MIX
Paper | Supporting responsible forestry
FSC® C016245

1% FOR THE PLANET
MEMBER

**Dedication**

To those who see a story of hope in every seed.

And to Mom, Dad, Kyle, Annie, Harper, and Griffin, who've helped me grow.

**Introduction** 6
A Regenerative Journey

## Part I: Reflect

1 **Plant the World You Want to Live In** 18
Viola, Wisconsin

2 **The Heart of the World** 44
Sierra Nevada de Santa Marta, Colombia

3 **Root and Plow** 74
Corn Belt, USA

## Part II: Rethink

4 **No Soil at the End of the World** 108
Kenya

5 **Eating the Rainforest** 146
Amazon River Basin, Brazil

6 **Hawaiian Pride B 4 Pesticides** 184
Kaua'i

## Part III: Regenerate

7 **River of Life** 228
Bluffton, Georgia

8 **Back to the Land** 262
Small Farms, USA

9 **Downstream People** 306
Mississinewa River, Indiana

**Other Farms** 342
**Resources** 358
**Acknowledgments** 359
**Index** 364

Welcome to an organic farm on the island of Kaua'i, founded after residents stood up against some of the largest agrochemical companies on Earth. KELSEY TIMMERMAN

# Introduction

**A Regenerative Journey**

Like a bushel of corn, my roots are in rural America.

Dad learned to farm while working on Grandpa's fields. I played in those fields. I made worlds in the corn, letting my imagination take me to faraway realms beyond the small part of the planet I knew. I was a warrior with a wooden sword fighting trolls and goblins, a superhero thwarting supervillains with my super-strength, and a kid exploring alien worlds carrying a walkie-talkie and a BB gun. But, as a member of the first generation of Timmermans not to grow up in a family that made a living off the land, I never dreamed of being a farmer. I never pretended to farm, because farming meant staying.

Instead, as soon as I was old enough, I left. I worked as a scuba instructor in Key West and Baja, Mexico. I hitchhiked from trailhead to trailhead in New Zealand. I immersed myself in lands and cultures beyond the monocultures of my childhood. For almost two decades working as a journalist, I've traveled from Australia to Zambia to meet people and write about how they live. I've especially focused on the farmers struggling to feed their families.

Travel confirmed that I belonged in rural areas. Whether in Africa or Central America, I found a familiar neighborliness. When I visited villages in Colombia or Nepal, I'd meet my new friends' grandmas and grandpas, aunts and uncles, and everyone else in the villages seemed to be cousins. I could relate. My teachers taught my parents. My wife's grandmother knew my grandparents. My distant cousin was the school superintendent who signed my wife's grandfather's elementary diploma. Around the world I was taken in and treated like family.

I'm never comfortable if I can't see the stars clearly. In 2007, my wife and I moved to Muncie, Indiana, onto a quarter acre at the

Twilight falls on Lake General Carrera near Puerto Guadal, Chile, where a regenerative farm is helping residents remember how to feed themselves. CLIFF RITCHEY

end of a cul-de-sac in a housing development. I suppose it was a quiet piece of suburbia to most people, but it was the big city to me. My first snowfall there was unsettling. The city lights reflected off the white ground, casting our neighborhood in frozen sepia, neither night nor day. Quiet didn't exist. There was always traffic, always alarms. I want the chirps of frogs in the spring and crickets at the end of summer, space enough for my kids to hit home runs into fields, wild berries, cherry tomatoes from the garden, catching toads, and lightning bugs. I like the old stories of friends and family coming together for the harvest or to raise a barn. I prefer sycamores as tall as buildings. Trees with stories. Keep the corporate ladder; give me branches to climb.

As much as I felt the connections with the land and the people in rural areas, there was no avoiding that so many of those connections have been broken. Farming once connected neighbors and communities, but as machines, oil, and chemicals have replaced people, as agriculture has become more industrialized, centralized, and commoditized, we've become separated from one another and from the natural world.

Rural lands and rivers and people are mined, poisoned, and exploited for profit at a great loss. Corporations farm farmers. Rural Americans lack access to sufficient healthcare and suffer higher rates of addiction, poor mental health, and a higher suicide rate than those who live in cities. Despite this, after a few years my wife and I left the city behind and moved with our two kids onto twenty acres on an Indiana country road that never earned a name. Once again, my neighbors were farmers and monocultures of corn.

As I started to plant roots in a new rural community, I grew more entangled in the stakes of farming. There are times, after

# Introduction

chicken manure has been spread on the fields, when it's hard—and unwise—to breathe outside. My mom has a nodule in her lung to show for the time she inhaled deeply after the field near her home was treated. I worry about my kids.

And then of course there are the other substances. Weed killer from a nearby farm drifted onto a friend who is an organic farmer. He's had stomach issues ever since. My son and I had to cut short a walk around our pond and hustle home when a neighbor began spraying his adjacent field. Another friend who farms corn and soybeans believes his dad and his uncle, both of whom died of a rare type of cancer, were killed by the chemicals they applied to their fields.

Modern industrial agriculture, practiced by my friends and family and neighbors to feed us, exposes them and all of us to cancer-causing chemicals and antibiotic-resistant bacteria. Agriculture contributes 31 percent of human-made global greenhouse gas emissions. It depletes our soil and uses more than 70 percent of our fresh water. American agriculture uses three pounds of toxic chemicals per person each year to produce vegetables and fruits that are significantly less nutritious than the food my parents ate growing up. Industrial agriculture treats earth like dirt.

Recent research is introducing us to a deeper understanding of soil, which is so much more than a collection of minerals. Soil is alive, and healthy soil serves a critical role as Earth's digestive system. Soil scientist David Montgomery writes that soil is the "ultimate strategic resource. For there is no substitute [for soil] as there is for oil, and it cannot be distilled, as fresh water can be from seawater, nor cleaned by filters as air can." The USDA estimates that it takes five hundred years to build an inch of soil, and yet the most common type of agriculture flushes it down our rivers into the ocean at such a rate—a dump truck load per second—that in 2014 a United Nations official expressed the fear that we have fewer than sixty years of growing seasons left. That's a pretty terrifying prospect, considering that 95 percent of the food we eat comes from the soil.

Industrial agriculture is changing our lives, communities, and planet, and my family had moved right back into the heart of it. As I wrestled with the risks of living near conventional farms, and climate change had me staying up at night worrying about the world we're leaving our kids, I began to hear about regenerative agriculture. I was learning that if farmers and ranchers followed certain practices, plants and soil could absorb enough carbon to mitigate and even reverse climate change. Maybe we weren't doomed? Maybe, during a time in which the climate news grows bleaker by the day, there was something more we—I—could do than hold a sign protesting climate change. Maybe there is hope? But there's so much more. Like many when they first learn of regenerative agriculture, I was so distracted by the hope of carbon sequestration I had a sort of carbon tunnel vision and couldn't see all of its other social and environmental benefits.

I became obsessed. I'm a guy who asks a question about something I don't know much about and then travels around the world looking for answers. "Who made my underwear?" took me to Bangladesh. "Who grew the cacao in my chocolate?" took me to West Africa. I've pursued my curiosity for thousands of miles—leading me to stand on glaciers and volcanoes, swim in the Amazon, and protect cows from lions on the Kenyan savanna—but no question has changed my daily life more

## Introduction

than the one that inspired this book: "What is regenerative agriculture?"

On the home front, it's meant that I inherited chickens, started a no-till garden, and built and rebuilt and rebuilt again a beaver dam by hand. I even became a shepherd. My four sheep graze our yard, only sometimes escaping to eat my wife's flowers. Occasionally the county sheriff knocks on the door at 10:30 pm to inform me that the sheep are across the road. Each morning after letting the chickens out, I evaluate the grass to decide where I will rotate the sheep next. Outside my front door, the world looks and feels completely different. I feel different. Because of my regenerative journey and the farmers I've met, the world is bigger and I'm more connected to it.

So what is this thing I've been obsessing about? Regenerative agriculture is much more than chickens and sheep and no-till gardening, but it's incredibly hard to pin down. Its spectrum of definitions reflects the unique qualities of the wildly diverse people who practice it in many different ways all over the world.

Indigenous cultures, individual farmers, communities, and nonprofits have their own understandings and definitions of regenerative agriculture, from a few loose principles to certifications with nearly a hundred specific standards to age-old practices that value giving more than taking. The giant agrochemical conglomerates—arguably the drivers and profiteers of the degeneration that spurred this movement—even promote their own very narrow definition, which serves their own purposes.

One of the most useful definitions came from a Stetson-wearing farmer in Georgia who told me that regenerative agriculture "restarts the cycles of nature." This was echoed by a spiritual leader of the Arhuaco people in the mountains of Colombia, who wore a hand-sewn white cap that resembled a snowy peak and told me that humans "have broken the rules of the earth, and its perfect cycle." Different continents, different cultures, different hats, but the same message.

My friend Michael O'Donnell, an organic farmer, told me, "I know regenerative agriculture when I see it. When I feel it." That wasn't exactly helpful to me as I struggled to understand, but it did encourage me to look and feel (and smell and taste) as I explored.

This is a book of experiences, stories of meeting farmers across the United States and around the globe who are working with nature and restarting its cycles. Many of them reject the destruction of industrial agriculture while embracing nature's abundance. But destruction and abundance appear side by side. It's precisely this contrast of people making a better future for their communities amid and in spite of the surrounding destruction that made this journey, to my surprise, overwhelmingly hopeful. I met a community in Kaua'i that rejected agrochemical destruction and embraced the abundance of their ancestors. I met traditional and Indigenous farmers fighting to hold onto their lands and culture, including the Maasai in Kenya and the Tupinambá in Brazil. I met farmers preparing for the end of the world and those trying to prevent it. I met a cowboy who wants to be composted and a young woman doing everything she can to convince the farmers in her community that their soil was worth more than the sand beneath it.

The farmers I met operate ranches, market gardens, and food forests. They are guided by diverse philosophies and traditions,

## Introduction

including Indigenous and traditional knowledge, permaculture, agroforestry, holistic management, and agroecology. As varied as they are in geography and culture, by the time I finished my journey I finally was able to discern a collection of common ideals that guide their work, from the way they treat the soil, grow a whole mess of things at once, and work with animals, to their demand that humans, who've been the source of so much destruction, jump-start the healing. This last part especially gives me hope. The doomers and some science fiction writers might believe humanity is the virus that needs to be wiped out, but regeneration recognizes that we are also the cure.

There's this idea that regenerative agriculture is hip and cool, marketable and new. As of the last few years, shoppers can buy regenerative crackers, cereal, popcorn, and beef. Headlines promote how regenerative agriculture can "save the planet" and one 2020 Democratic presidential candidate, Tom Steyer, made regenerative agriculture a critical part of his platform to fight climate change. Actors Rosario Dawson and Woody Harrelson each narrated a film on regenerative agriculture.

But as I traveled and met regenerative farmers, I discovered something that often gets lost in the hype: Regenerative agriculture isn't an exciting new thing. Regenerative agriculture is an exciting old thing. As it's been practiced by Indigenous people for thousands of years, it is not centered on extraction but on relationships with the natural world; it is not just about farming, but about our relationships with science and technology, each other, and ourselves. While it's grounded in the soil, it's about human rights and environmental justice, about diversity versus monocultures, the pursuit of profit versus life. I started to see that thinking regeneratively is about so much more than growing food or adopting certain farming practices. It's something each of us can practice in our daily lives. It's about reconnecting with the natural world during a time when many of us prefer to stream shows from the cloud than watch the clouds from a stream.

Every one of us has a farmer in our family tree. If you aren't a farmer yourself, at some point, one of your ancestors decided to make a living doing something other than farming. In 1862, when President Lincoln signed the legislation that created the Department of Agriculture, he called it the "people's department" because 90 percent of Americans were farmers. By the time my grandfather was born in the 1920s, only 30 percent of Americans were farmers. Today, fewer than .1 percent of Americans live on commercially viable farms.

It's not easy being a farmer. Fifty percent have lost money since 2013, and many must rely on imported labor to function. Nearly three-fourths of farmworkers in the United States are immigrants who left their homelands to work someone else's. Extreme poverty is mainly a rural phenomenon. Nearly 80 percent of the world's poor live in rural areas where they wrestle with invisible forces that are often slow and impersonal. Industrial agriculture keeps them hungry as it also erodes their communities.

My regenerative journey helped me actually see some of these people—not something we are able to do very often—and also make visible the forces that impact their lives and ours, and how our futures are intertwined. Rural communities, our soils, our relationships with nature all need to be regenerated.

It's hard for me to admit, but before this project, I had given up on farming. It was the last place I thought would offer answers to the biggest environmental and social issues of our time. But on

**Introduction**

my global regenerative journey over the last five years, I've started to see that our hunger, and the agriculture it requires, can be the gifts that connect us to chloroplasts, lions, mycorrhizal fungi, our fellow humans around the block and across the world, and our responsibility to play an active part in a regenerative future.

I learned that not all humans are contributing to the degeneration of our planet. In fact, some are proving that a different relationship with the web of life is possible. One that is positive. One that regenerates. One that is renewing Earth.

---

*Land Acknowledgment*

*Building mutual relationships with Indigenous peoples is part of our work to restore Earth, the home we all share. Patagonia's headquarters is located on the unceded homelands of the Chumash people in what is now known as Ventura, California. Because the people in this book are in locations around the planet, we acknowledge the many Indigenous communities who have stewarded the lands and waters of each of these places since time immemorial. We are also grateful for their continued leadership in the environmental and climate movement today.*

---

# Part I: Reflect

# Chapter 1
# Plant the World You Want to Live In

# Viola, Wisconsin

**Regenerative Agriculture Is Permanent Agriculture**

I knew I was in the right place. I could tell by the weeds growing next to the building. On the neighboring farms of conventional corn, nature was forced into straight lines, but this place was a thriving chaos. If I hadn't known that the 106-acre farm behind the building was one of the most ambitious regenerative farms in the United States, I would've thought it was abandoned. But it wasn't abandoned—it was abundant.

The neighbors grew one crop: #2 yellow dent corn, which is inedible to humans unless processed. This place, New Forest Farm, produced hazelnuts, chestnuts, walnuts, apples, pears, plums, cherries, grapes, persimmons, blackberries, raspberries, blueberries, elderberries, currants, gooseberries, hops, grapes, kiwis, basil, chives, dill, mint, oregano, parsley, rosemary, sage, and thyme, asparagus, beets, broccoli, cabbage, carrots, celery, cucumbers, eggplant, lettuce, onions, potatoes, pumpkins, squash, tomatoes, and corn that humans could actually eat. The farm was also home to chickens, pigs, sheep, goats, at times cows, and a tall, serious-faced man dressed like a park ranger,

PREVIOUS SPREAD: Mark Shepard, whose 106-acre farm is a permaculture oasis among fields of corn, guides Kelsey through his orchard, where he practices his STUN method—Strategic Total Utter Neglect. CLIFF RITCHEY

OPPOSITE: At least eighty-seven species of birds, including this Baltimore oriole, call New Forest Farm home. Mark, working with the plants and animals on his farm, brought this land back to life after it had been farmed to death. CLIFF RITCHEY

who lumbered out of the building, took one look at me, and demanded, "What the fuck do you want?"

For a moment, I wasn't sure. But then I considered firing the question right back at him. It would've been appropriate because what this man, Mark Shepard, wanted—and created—is what I was there to see. But I chose to laugh it off and then complimented him on how alive this place was and for writing a wonderful book, which gave me a glimpse of what this tour would be: lessons in history, philosophy, genetics, biology, geology, and social criticism.

Mark, the pioneering author of *Restoration Agriculture*, had moved onto this hilly land in southwestern Wisconsin in 1996 and had asked it the same question: "What do you want?" At the time, two-thirds of the farm—"small, stupid little fields"—had once grown corn, and the rest was a pasture grazed down to the height of a golf course. The land was eroded and gullied. It had been farmed to death. With a lot of work, Mark has restored it to how he imagined it once had been. Mark, and many others who've invited him to teach around the country and the world, believe that what he had accomplished was truly a regenerative agriculture. And Mark thought regenerative agriculture is the key to solving the world's food and climate crises.

Perhaps more than any I visited, Mark's New Forest Farm offered the best introduction for me as I tried to come up with a definition for regenerative agriculture. In his own way, he demonstrated all of what I ultimately decided are the beliefs regenerative farmers have in common, including the conviction that humanity can be a force for good. Although, regarding this critical point, Mark seemed a little grumpy, like he might not like humans all that much. I suppose

## Plant the World You Want to Live In

working on regeneration in a world that lives off destruction will do that to a fella.

I also wanted to know if it's really possible to farm in the Midwest and make a living without growing corn or soybeans.

Mark welcomed me into a building, a former cidery open to the public that had closed after a state law prevented small farmers from selling cider directly to consumers. Now there was a bed (for guests, maybe?) tucked in a corner, a battery charger, a poster promoting three different varieties of Shepard's Cyder. Another poster featured the Christian works of mercy: "Feed the hungry, clothe the naked, give drink to the thirsty, shelter the homeless, visit the imprisoned, care for the sick, and bury the dead." A very dead and very unburied stuffed bobcat lounged on a trash bag atop a table. There were musical instruments everywhere: guitars, banjos, and even a baritone.

At Mark's suggestion, I filled my water bottle and applied sunscreen. He hooked a large finger swollen from years of work through a gallon milk jug of water, and led me from the building into an oasis of food.

We walked onto a relatively flat area at the top of a valley that sloped down fifty feet. From a drone's point of view, the flat area would look like a giant had bent over and touched the ground, leaving a circular fingerprint that was 175 feet in diameter and swirling with green loops hugging the slopes like contour lines. From within we were lost in a maze of earthy-smelling hazelnuts. Hazelnuts don't naturally grow in mazes. Mark had made his impression.

We worked our way through the maze, and chestnut branches brushed our shoulders and cheeks. Stopping in a grassy alley

Mark walks through the maze of life, past the chestnut trees in the overstory and through the "weeds" that thrive below, observing the trees, developing and explaining his version of AI—Actual Intelligence. CLIFF RITCHEY

between the trees, he pulled down his wide-brimmed camo hat, as weathered as a scarecrow's, slugged back some water, and sighed. He was recovering from a recent flare of Lyme disease.

"I'm a little bit sluggish right now," he said. But he assured me disease-carrying ticks were far less of a concern now that he'd incorporated guinea fowl on the property. Guineas eat ticks. And Mark can eat guineas.

Among the farmers I'd meet on my journey, working with animals was going to be a theme, as was managing one kind of life by incorporating a diversity of other kinds of life. This strategy is the opposite of what's employed by the row-crop farmers who were Mark's neighbors, and for that matter, my neighbors back in Indiana. They relied on the "'cides"—'*cides*, of course, meaning killers. Pesticides, the killer of pests; insecticides, pesticides that target insects; and herbicides, the murderer of undesired vegetation or, as some might call them, weeds.

What is a weed, anyway? A plant we view as not belonging and undesired—as if plants are restricted to existing where we want them, while we belong everywhere.

The 'cides are technologies developed for or used in the service of war. Their precursors include the infamous chemical cocktails Agent Orange and Zyklon B, which the Nazis used in their gas chambers. The process that created the ammonium nitrate Mark's neighbors applied as fertilizers was first used to make nitrogen bombs. Chemical killers of people were and are now deployed to grow food and destroy all things except the GMO crop a farmer wants to keep alive. GMO stands for genetically modified organism, sometimes also referred to as bioengineered or GE—genetically engineered. People have a lot of opinions about

GMOs, the majority of which are crops modified by agrochemical companies to resist the killing chemicals they just happen to sell.

As we tucked into the shade of a wall of chestnut trees, Mark became more energized. "All of this right here was once upon a time oak savanna. Europeans came here, cut it all down, burned it all up, then plowed it and planted it to corn and beans and whatever else. It all washed away."

And then modern farmers moved in, growing corn and soybeans exclusively, which together account for more than half of the cash crops grown in the United States. These are annual crops, planted and harvested every year, and farmers rely on 'cides or tilling to manage unwanted pests and plants. After harvesting, for half the year, the fields sit as bare, often lifeless deserts, gray and brown even in the springtime when the rest of the world grows green.

But Mark's green hillside no longer needs to be tilled or replanted. The trees and grassy alleys hold the living soil, preventing it from washing or blowing away. Providing sugars and carbon to the soil and being rewarded with nutrients, the roots allow the soil to store more water, reducing the impact of heavy rainfalls and droughts.

Mark pulled on a thin chestnut branch weighted with green, spiky burrs. "We get the equivalent of a soybean right here," he said, showing me the healthy crop of nuts the size of golf balls that, like soybeans, could provide protein to our diets. "It's perennial, produces three times the oil. And the shells burn as hot as anthracite coal. Every single year you can cut this wood, a fuel if you want to use it as a fuel or as a mushroom substrate. We can grow our 'soybeans' on a tree, site-adapted, no fertilizers, and we can use animals as our fertilizer and weed control."

Mark designed the farm based on how water flows across it. The fingerprint contours, created with swales of earth, sloped slightly down. We were following in the water's footsteps, switchbacking at the end of an alley and then turning onto the alley beneath it.

Mark abruptly stopped his lecture and stared at a tree.

"Who are you? Show yourself!" he demanded as if spotting a lurker.

His body language begged for silence. I hadn't realized it until then, but we were surrounded by birds. Their wings beat overhead and shook nearby branches as they landed or took flight. Mark had been walking these rows for a quarter of a century, and now there was a bird call he couldn't identify. The moment passed and he didn't hear it any longer, but he did start interpreting other chirps, trills, and whistles, and he even talked back, assuring an alarmed red-winged blackbird that we would soon be moving on.

"This spring I started a catalog of all the birds that I saw with my naked eye," Mark said. "How many species do you think that I've seen so far?"

I thought about how silent the fields around my home are and answered Mark with a shrug of my shoulders.

"Eighty-seven," he said. "Some are migratory. You won't get that many out in the cornfields."

•

Mark himself had migrated from the suburbs of Boston, where he was raised. His father always kept a garden and was an early adopter of the type of composting promoted by Rodale's organic farming and gardening magazine. Hippies would come by to learn about it. However, Mark wasn't a big fan of working in the sun. He preferred the woods. One of his grandfathers was a backcountry guide in Maine and another worked on a dairy farm, living lives closer to the way Mark wanted. But out of college, Mark got a job as a mechanical engineer. He was paid well, had a forty-minute commute, wore a tie—and hated it. He quit and went back to school to study ecology. In 1986, Mark, with his newfound knowledge, wanted to try his hand at living off the land. He hitchhiked to Alaska to take advantage of the state's homesteading program, and claimed five acres 3,500 feet up a mountain 300 miles from the nearest town and five miles from a road. He lived there with his best friend, Jen, who became his wife.

He became interested in permaculture. *Permaculture* is one of those words I would hear around the world while on my regenerative journey. It's less a system of how to farm than it is an ethic, a holistic lens through which to see and make decisions, based on three interconnected pillars: Earth care, people care, and equitable distribution of resources among people and Earth. Through close observation and self-reflection, permaculture offers an approach to living in balance with nature. For example, an herb garden planted in a spiral is a classic permaculture technique, offering an efficient way to grow diverse plants while maximizing sunlight and water. Those herbs might go on a pizza baked in a mud oven constructed from materials readily available on the land. There are twelve permaculture design principles and the spiral and oven alone would meet several of them, including: observe and interact, catch and store energy, obtain a yield, produce no waste, value diversity, and use the margins.

Mark took a permaculture course and received his diploma signed by Bill Mollison, who actually coined the term in the 1970s. Before long, Mark began teaching permaculture courses himself.

The contours of the farm are designed around this wetland area. Mark asked the land what it wanted to be and is helping it return to a version of the oak savanna that once dominated the landscape.
CLIFF RITCHEY

But after eight years of homesteading, Mark and Jen were expecting a child and decided to move somewhere less remote. He still wanted to be self-sufficient and live off the grid, but he was beginning to think bigger. He felt that permaculture was often reduced to brick ovens and spiral gardens of herbs by people seeking to partially remove themselves from the dominant food system, which they felt was unjust and reliant on slavery and environmental degeneration. He agreed with them, but meanwhile the food system barely noticed their absence. Mark wanted to scale things up and prove that permaculture—not just on a few acres, but on more than a hundred acres—could live up to the original intent of the word: a permanent agriculture.

"I thought agriculture meant growing food," Mark said. "We are supposed to imitate nature and model natural ecosystems. So I said, 'All right, that's what we'll do. And we'll do it as an enterprise, instead of as a nonprofit, educational thing, or a little backyard demonstration site. It's got to actually feed people. It actually has to pay its own way in this economy.' My intention was to get staple food crops from a perennial system."

He wanted to feed not only his family, but his community and region, and show that permaculture at scale could feed the world justly.

So he and Jen planted permanent roots—hundreds of thousands of roots—in Viola, Wisconsin, on land degraded enough that they could afford it.

He asked the land what it wanted to be. And he learned that what it wanted was to return to what it had been before the invasion of the corn growers.

## Plant the World You Want to Live In

He told me about the biome that originally thrived in this area—oak savanna. He described the towering trees (oak, chestnut, beech) in the overstory, and beneath them the layers of life that provided food for humans and animals all the way down to the mushrooms bubbling to the surface from under the soil. By contrast, of course, conventional agriculture focuses only on promoting and harvesting from a single layer.

Mark began a lecture that spanned tens of thousands of years of geology, his hands becoming glaciers shaping the landscape.

Next thing I knew, his arms became the swinging trunk of a mastodon. Mastodons and bison helped manage the savanna forest, preventing it from becoming too dense and dominated by the overstory. Mastodons would tear bark off with their tusks, devour low-hanging fruit, and then push trees over to get at the rest of it. Giant herds of bison would take days to pass, all the while grazing, trampling, dropping their nutrient-dense waste, and contributing to and shaping the landscape.

Getting a feel for the way this man thinks, I asked, "Are you about to tell me that you're going to get an elephant?"

Mark didn't dismiss the idea. In fact, he had put some thought into it. There's a short elephant in Indonesia he'd cross with a breed from India, which has genetics closest to a mastodon. He'd select for small, hairy beasts that could work in the contours and survive Wisconsin winters.

Elephant-less, Mark planted an ancient oak savanna forest and its layers on what had been a single-layer cornfield. He described the layers, top to bottom, as we walked beneath them. He started with the chestnuts in the overstory and then went down to the food- and income-producing species that thrived below. From tallest to shortest there were cherries, plums, and apricots, hazelnuts, raspberries and blackberries, grapes, gooseberries and currants, and the green grass and fallen fruits and nuts on the forest floor, where they could feed pigs, cattle, sheep, and other grazing animals kept in place by mobile electric fencing.

Mark chose chestnuts for the primary overstory trees instead of oaks, which take decades to produce acorns. Novelist Richard Powers wrote that the chestnut was "the backbone of entire rural economies" in the early 1900s. The chestnut was among the fastest-growing, tallest, and largest trees in the forests. People would throw rocks at the trunk, and nuts would fall by the shovelful. Railroad cars overflowed with chestnuts used to make flour, bread, and candy, and roasted over open fires during the holiday season. The wood was used for everything from cribs to cabins to caskets, railroad ties, and telephone poles. Powers calls it "America's perfect tree."

One problem: chestnut blight, caused by a fungus that hitched a ride here on a Chinese chestnut at the turn of the twentieth century. In only a few decades the fungus would wipe out most of the mature native trees in the United States, which once numbered four billion. Chinese chestnuts were resistant to the fungus but, unable to handle the cold of Wisconsin, couldn't effectively step in and take their place.

But Mark is working on it. He uses chestnuts that are a hybrid of the two. His strategy is to plant way more than he needs, knowing that many will succumb to the blight and others to winter. He simply concludes that the ones that get sick or die aren't suitable for his location; they lack the proper traits. Instead of trying to keep

unfit trees alive, he cuts them down to burn as fuel or to inoculate to grow mushrooms. Even dead, the roots would contribute their nutrients to their fit neighbors. Any branches left behind would decompose and become soil.

Methodically, Mark selected for the trees that survived. Experts told him that he should apply for a grant, start a study. One plan promoted by academic horticulturalists required a six-figure investment per acre over five years before a single chestnut would be harvested. Mark didn't have the patience or the resources for that. He planted and selected and observed and studied and produced all at the same time. Within three years, he was harvesting chestnuts.

And way down in some of the alleys between the trees, he planted asparagus. He approached the asparagus like he approached everything: Plant a lot and don't micromanage it.

With mastodons and bison gone from the landscape, Mark manages his forest with a chainsaw to keep the overstory from becoming too dense. CLIFF RITCHEY

Regardless of the layer, observing the life of each plant and the system as a whole are critical to his approach. He asked me to think about the life in the ditches by the side of the road. No one waters or sprays them, and the ditch plants are often more diverse and productive than the fields of annual crops next to them.

Mark's methods don't require much of his time or money. He knows that traditional orchardists would be appalled that his apple trees aren't perfectly pruned and in tidy rows, but even one salable apple means a profit for him since his costs to produce it are next to nothing. While other orchardists spray and prune the trees and pick up the fallen fruit, he leaves the trees alone and moves the pigs in to feast on the bad apples, making them fat and happy and removing rotting fruit, which is a source of disease. The trees fit for his farm live; the others die. Nature is quite capable of managing herself with an occasional assist from Mark.

"I'm just trying to focus on ecosystem health," he told me. "Because when you start to restore Earth's ecosystem, part of what intact ecosystems do is create soil. They take rocks and water and air, and this magic of life actually creates more fertile soil, more biodiverse habitats. This planet trends toward maximum richness and abundance and diversity and stability over time, no matter what the cosmos can throw at it. And until we imitate that, we're fools."

"Sometimes we think we can remove ourselves from natural forces," I said.

"And humans have been living on that technological pipe dream for two hundred years," Mark said.

He called his method of overplanting and ignoring "the STUN method." STUN originally stood for "Sheer Total Utter Neglect," but sometimes, because his acronym has been used as a criticism of his methods, he switches the "sheer" to "strategic" to make it sound more intentional.

"We don't need to have these beautiful, grand, perfect, spotless, gorgeous foods," Mark said. "Nature's a mess. Nature's very productive. And it's very low cost. I may get half the yield of conventional growers, who would ask me, 'How can you pay the bills?' But if they're getting a full yield and they've got all these expenses… their net is…. " Mark pointed down. "And then I get half a yield and my expenses are…." He made a zero with his hand. "Who has more net, less work? Because all of this is not only about expenses, but it's also about labor, time, and effort. Screw that. I'm going fishing!"

Annual grains consume much of our land, diets, and imaginations. My farmer friends and neighbors will often remind me they feed the world; that industrial agriculture, and all of its synthetic inputs—fertilizer bombs and payloads of 'cides—are necessary to meet the needs of the world's growing population. If we focus only on gross yield and not net calories, they are correct. But industrial agriculture requires the burning of a lot of fossil fuels. For instance, a gallon of gas contains about 31,000 kilocalories. So when the energy for the tractor passes, grain trucks rumbling on country roads, production and hauling of chemical inputs, all the transportation to take feed to livestock, and the required processing to make corn products edible to humans are deducted, Mark figures his farm—all of the berries, nuts, veggies, and animals—actually produces 2.6 million more human-available calories than if it were all a field of corn. Rejection of industrial agriculture practices, that was something I'd hear from many farmers on this journey.

We were winding our way up out of the valley in an area that ran outside the contour fingerprints and was more exposed to the sun when we heard snorting. Mark pointed to the grass.

"There's a little piggy shit," he said, as floppy-eared hogs of various sizes trotted our way. "Everyone has a name," he told me. "In order for us to maintain our humanity, I feel that we have to treat our fellow beings with kindness and love and respect and all that kind of stuff. They'll come up and they might want some scratches and rubs, want to talk to you."

"Hey guys," Mark said in a higher-pitched, piggy-talk voice. "Let's go over into these hickories where it's cooler.... Oh, you guys have been swimming."

The pigs nipped at our feet and brushed up against us as we led them into the shade provided by a grove of hickory trees that smelled slightly like butterscotch. Mark showed me how to gently push them away. A pig named Sour Diesel Bud flopped on his back and Mark rubbed his belly. Petunia, Iris, Hawkweed, Dahlia, and Begonia looked on. These particular pigs weren't his. He runs a sort of pig rehab. A pork producer sends Mark the pigs that aren't going to make weight. At Mark's farm, they eat grass and bad apples, and dispose of any leftover produce, such as acorn squash, asparagus, and green bell peppers that he grows.

Mark kept telling me that he doesn't have to work hard, but he knew and paid attention to every plant on his land—which sounded overwhelming to me—so I had to wonder. He does get some help here and there, especially when it comes to harvesting.

## Plant the World You Want to Live In

But he doesn't welcome interns. For him, "intern" sits right next to "internment" in the dictionary and, he says, "We made that illegal in 1865." Mark prefers partnering with other farmers. If someone wants to come onto the farm to collaborate on projects, Mark is happy to welcome them, like with the pigs, and provide guidance.

He offers his knowledge and encouragement to prospective farmers, too, with a dose of reality. "You want to be a farmer?" he asks them. "Okay, come on out. I'll be here to help coach you along the way. I'll help you with setting up the business, absolutely everything. But you've got to figure out how to finance it. If you can't do that, you don't have what it takes."

The numbers have worked for Mark. He bought a property. Improved the property; the value went up. Borrowed against that property and did the same thing again. He's repeated that for twenty-five years. In addition, he's layered on several business ventures, including a nursery selling his hybrid chestnuts and hazelnuts and a company that designs food forests like his. Plus, he consults on farms around the world. But he never forgets what it takes to pay his mortgage: five steer, a dozen pigs, a few acres of chestnuts, hazelnuts, and asparagus. And as someone who wants to be prepared for any circumstance, he still maintains his commercial driving license just in case he needs to take a trucking job in a pinch.

Mark whistled, communicating to the pigs, who had left our scratches to go back to their swimming hole. "There's mulberries!" he hollered. As they came running back, a Baltimore oriole nestling fell from a tree. Mark encouraged the little bird as it found a hiding place where the pigs wouldn't eat it.

While the pigs will eat pretty much anything in their path—they act as mowers and brush clearers as well as vacuum

cleaners on the farm—he puts rings in their sensitive noses to keep them from rooting into the soil too much. I tell Mark about a brush problem I'm having in my woods and pond. Autumn olive and bush honeysuckle are taking over and choking out the woods, and the pond banks are growing over with willow.

Mark had an answer for all my plant problems: "Have a relationship with them."

"You've got to understand your place has been fucked over for two hundred years," he said. "Yeah, it's nothing like it could be, should be, would be... nothing. It's wasteland."

Curious and friendly pigs browse the understory and eat any fallen fruit before it attracts harmful pests. Mark calls each of them by name and says that all beings should be treated with love and respect. CLIFF RITCHEY

**Plant the World You Want to Live In**

*A wasteland?* That stung. It's my favorite place on the planet. Sure, it's a little swampy, parts of it had been clear-cut in the not-so-distant past, but it's home to some of my favorite trees, animals, and people. But then I thought of Mark stepping onto his property for the first time. He wouldn't be seeing the destruction, but instead imagining the regeneration. What if I did the same? What if I asked the land what it wanted?

He tells me that I could harvest the willow and use it to make baskets. It's one of the first plants to have pollen in the spring, so it's good for the bees. Willow tea helps treat headaches and fight Lyme disease. I could use a handyman jack to break the roots of the honeysuckle and autumn olive. I could let the plants biodegrade in place or chip them to add to my compost pile.

At the top of the hill we came to Mark's house. It was practical and designed for efficiency. He stopped, turned toward me. His tone shifted. Earnest and slightly defensive, he said, "I'm not saying that people have to live the way I live, but this is an agricultural model that works. I haven't paid an electric bill since 1980."

I was surprised by Mark's tone. It was almost threatening. I think Mark worried what I'd write if I looked at his house and his life through the eyes of the average American. He has created a place teeming with life and is the most self-sufficient person I've ever met, yet most Americans wouldn't want to do the work it requires to live as he does. Many of us, me included, talk about living sustainable, climate-friendly lives, but when we meet someone like Mark, all the recycling, electric car driving, farmers marketing, carbon offsetting, and ethical consuming pales in comparison, and, honestly, makes me uncomfortable as I ask myself: "Am I willing to change my lifestyle?"

Mark had solar panels for electricity and hot water. There was some type of home cooling system involving water that ran beneath the house. His chicken coop, his compost pile, his garden, his logs of mushrooms—it was as if he had applied every permaculture design principle and trick. His garden had all the veggies he and his family could eat, in addition to 139 medicinal plants.

There wasn't a driveway to the house, a half-mile walk from the parking area. In the winter, snowshoes are sometimes required. Everything the family bought is carried or pulled in a wagon or on a sled. Mark said that no matter what had happened since you'd left the house, the walk back gave you time to sort it out. They didn't use flashlights, just their night vision, the moon, and the stars.

That evening I joined Mark, his father, and a college student named Sammy for dinner. We sipped gin and tonics while Mark cooked bratwurst over a fire pit behind the house. Later, we ate blueberry pie made with berries I had helped pick.

Mark and Sammy recounted their recent experience in Tanzania with an organization called Mainsprings. Mark has worked on more than a hundred farming projects around the world, and this is among his favorites. Originally an orphanage, Mainsprings now serves as a training center focusing on education, health, and permaculture. Before Mark, their farming program was limited to monocrops. Now they produce nearly a ton of fruits and vegetables a week—from papayas and mangoes to cucumbers and carrots—enough to feed their four hundred students, with some left over for the surrounding community. Mark said the leaders there were skeptical of his methods at first, but they couldn't argue with the results.

## Plant the World You Want to Live In

"When the whole COVID thing happened," Mark said, "they just shut the gate. They were producing all of their own food."

•

Food is survival. It may be more apparent in Tanzania than in the United States, but who can forget how our shelves went bare during COVID? Our just-in-time food system nearly collapsed amid horrifying stories of workers dying after being forced to endure unsafe conditions, hogs being suffocated, slaughtered, and burned when processing plants closed, and tankers of milk being dumped on the ground. All of this happening at the same time food pantries saw the longest lines in memory. But Mark was prepared for whatever, and, like most of our society, I was not.

The day before Indiana passed the statewide lockdown in 2020, I bought cans of soup in flavors I'd never eaten before—chicken wonton, chicken gumbo. Most of the meat was gone. And, of course, I grabbed every roll of toilet paper I could find, even though we already had our largest-ever supply at home.

Our food system barely held together during a crisis, but even when times are good it provides some of us with cheap, unhealthy food, while still leaving more than one in ten people undernourished. It's not good for the planet or people and it's not resilient. I know this intellectually, most of us do, but we don't live like it.

While reading Richard Powers's novel *The Overstory*, I came across a simple set of questions that reveal this disconnect. They are asked by a character who's an environmental activist:

"Do you believe that humans are using resources faster than the world can replace them?"

Mark's house sits a half-mile walk from where he parks his car. His productive garden provides him with food and also contains 139 kinds of medicinal plants. CLIFF RITCHEY

"Do you believe that the rate we use them is increasing?"

Regardless of political ideology, I think we all answer yes to both questions. I certainly do. I can add my own simple questions that are equally revealing:

Are we depleting soil faster than it takes to build it? *Yes.*

Is the climate changing? *Yes.*

Is our food system resilient? *No.*

If these simple questions give you anxiety like they do me, find comfort that there are people like Mark who provide answers to the problems they reveal.

However, Mark's critics accuse him of being impractical. They argue that his system could never scale to feed the world. That he's out of touch with reality. Yet most of us still trip through existence completely reliant on systems built for endless growth that rely on finite resources. Who's the one out of touch?

"I was wondering if I could buy a ticket?" I asked Mark. "So in case of an emergency I could come live at your house?"

It was no surprise that he shut me down immediately. His activism speaks loudly: Plant the world you want to live in.

•

I couldn't get over how Mark kept the incredibly complex system he'd created in his head. He didn't take notes. There wasn't a spreadsheet to reference or an app to guide him. Walking around and noticing was what his work looked like most of the year.

## Plant the World You Want to Live In

"I don't see you scribble anything down. How do you keep track of everything?"

"Here's the technology I'm into: wild stuff," Mark said. "Here we go... observation exercise. Close your eyes and just listen for a minute. Listen to all of the information coming at you."

Birds. A rooster in the distance. The wind through the leaves of a chestnut and an oak. Maybe it was my imagination, but then suddenly I thought I could hear the slope of the hill above me and the openness of the valley bottom before me.

"How much of the audible spectrum are you able to notice?" he asked.

I wasn't sure what the answer would even look like. It would have to be measured in something grander than decibels.

"All of it," he said, moving his eyes around, twisting his hands, and breathing deep to signify using all of his senses. "So there's this hemisphere around us... an infinite number of points around a central point. How much of that stuff was useful information that you could base decisions on? All of it. Everything was telling you something. So we're not even paying attention to the planet we fucking live on. That's AI. That's Actual Intelligence. We're not even paying attention."

Actual Intelligence. I was starting to see why the first principle of permaculture is to observe and interact. If we are going to care for the Earth, its richness and abundance, we have to hear, smell, taste, touch, see, experience, and pay attention to it first.

"What if we did?" I asked, unsure if it was a rhetorical question.

"What if we did?" Mark responded.

# Part I: Reflect

## Chapter 2
## The Heart of the World

**Regenerative Agriculture Is Rooted in Indigenous Knowledge**

# Sierra Nevada de Santa Marta, Colombia

From where I stood, I saw heaven.

The white peaks hung in the distance, a world away from the lush, hot forest garden where I was staying. My mind always wanders to places I'm not. I imagined standing atop the mountain—my hosts referred to it as *chundua*, which means "heaven" in their language, Chibchan—snow crunching beneath my feet, squinting into the wind, looking down. If I could see myself from the perspective of heaven, what would I look like?

I had visited the Arhuaco, an Indigenous people who live in the mountains of northern Colombia, seven years ago to learn about their fair-trade coffee business. Things had been going well. The coffee-generated income had enabled them to address some of the medical, educational, and legal needs of their community. And there was room for growth. More money to be made. Their non-Indigenous partners wanted to improve the roads to make it easier to get the coffee to market—but the Arhuaco balked. That would have given the outside world too much access to their cultural and spiritual capital, Nabusimake. They were content; their needs were being met. They didn't let the growth-obsessed outsiders change them. They just wanted to preserve and sustain their homelands.

The Arhuaco believe their land of rivers and snowy peaks in a remote, unspoiled region of Colombia—home to more threatened endemic species than any other place on Earth—is the Heart of the World. CLIFF RITCHEY

At that time I had never heard the term "regenerative agriculture," but looking back I recognized that my first visit with the Arhuaco was my introduction to a truly regenerative way of life. I returned to learn more from them.

The Arhuaco believe that Mother Earth birthed them here in the Sierra Nevada de Santa Marta to be the caretakers of a region they know as the Heart of the World, the center of their universe. I found it on a map. The black border that defines the region is also shaped like a heart—not a Valentine's Day heart, but an anatomical heart. The aorta juts up the Caribbean coast, the ventricles push south into the mountains. Rivers capillary down from peaks in the center that touch the sky at nearly nineteen thousand feet.

The Heart of the World is home to more threatened endemic species than any other place on Earth. The blue-bearded helmetcrest hummingbird, identified by its supervillain-like black-and-blue mask, is so incredibly rare that it hadn't been seen by researchers for sixty-nine years before two scientists snapped a photo of it in these mountains in 2015. The starry night harlequin toad's black spots and white pattern look like the clear night skies above the mountains. These and other toads, viewed by scientists as key indicators of the health of an ecosystem, are the tiny authority to which the Arhuaco turn. The toads' songs help them decide when to plant their crops.

*Science* magazine has named the Sierra Nevada de Santa Marta the most irreplaceable site in the world for the conservation of protected species. And the Arhuaco are committed to protecting it. But their culture is also threatened, so much so that the United Nations Educational, Scientific and Cultural Organization

## The Heart of the World

(UNESCO) added the Arhuaco's "ancestral system of knowledge" to their "List of Intangible Cultural Heritage in Need of Urgent Safeguarding," stating that the "skills and sensitivity to communicate with the snow-capped peaks, connect with the knowledge of the rivers, and decipher the messages of nature... is believed to play a fundamental role in protecting the Sierra Nevada ecosystem."

The Arhuaco see the climate changing. The bees sound different. Plants are migrating to higher elevations. But the Arhuaco believe that their prayers, rituals, rites, dances, songs, and other such offerings to spiritual powers in the Heart of the World keep the entire world from falling into destruction and despair. They don't take anything more than they give back and have the smallest of material footprints, but believe they have the ultimate impact. This shows up in the intangible ways they interact with and think about the world, and also in their very tangible farming practices.

The Arhuaco minimize soil disturbance and control unwanted vegetation through mulching. They plant heritage sugarcane on small plots in the forest during a full moon and ask permission from the earth before harvesting. Their gardens overflow with an incredible variety of plants from bananas to cacao, confusing pests with the variety of colors and smells, and attracting beneficial insects and pollinators. Even their coffee is grown as part of a diverse agroforestry system. All planting and harvesting are done on the human scale by hand and machete, and, of course, no chemicals are used. I had hoped to see these practices firsthand—as I would on other farms around the world—but for my return visit to the Heart of the World, the Arhuaco had other plans.

Freddy directed my friend and photographer, Cliff, and me to take off our shoes and sit on a curved rock wall in the middle of the lush forest garden in the village of Pueblo Bello. Facing us, he was in a chair made of boulders, sitting on one and leaning back against another. His seat, glossy from wear, fit him perfectly.

Freddy and our guide/translator, Miguel, on his own boulder chair, both wore hand-sewn cotton tunics and pants and white hats, which they'd woven themselves. The hats, which

Freddy approaches a home on a hillside. The Arhuaco feel that their prayers, rituals, rites, dances, songs, and other such offerings to spiritual powers keep the entire world from falling into destruction and despair. CLIFF RITCHEY

represented the snowy peaks in the distance, indicated their role as Arhuaco spiritual leaders, mamos, living in harmony with nature and concentrating their life force on the life forces of all living things from the mountains down to the sea.

"This is your second time in these mountains. What made you come back?" Freddy asked. "Tell us, tell Nature."

Freddy held the fate of our journey in his hands. He would decide if we would have the opportunity to learn from him and other Arhuaco mamos or if we had come all the way here from Indiana just to turn around and go home.

I thought about my first visit, and how I had learned that the Arhuaco see themselves and other Indigenous people as the older siblings and the rest of us as children. And how I had felt like a child when I couldn't identify a cucumber leaf from a watermelon leaf. They had laughed in disbelief that someone like me surrounded by gadgets and the age of information wouldn't know something so fundamental. They had inspired me to plant my first garden when I got back home. They had predicted it would take two more visits for me to understand their lessons.

This was visit number two, and a lot hung on what I was about to say. I took off my Rural King hat and ran my fingers through my hair. I had purchased the hat from a farm store at home. At the time it seemed like a perfect choice: a local farm hat to wear on this journey to meet farmers around the world. But now, on the rocks surrounded by palm trees and heaven, "Rural King" was a label I wanted to run from. Sure, I could tell the difference between a cucumber and watermelon now, but I still knew so little about this kind of rural life, much less about royalty. The part of my gut reserved for pending embarrassments and cultural missteps began to tingle.

I took a deep breath and sat my sweat-stained hat behind the rock wall. I told Freddy about my grandpa, who farmed until he was eighty-five, and how his children sold the land and got out of farming. I told him how my home was surrounded by monocrops of corn and beans whose production ravaged the soil. I told him about my farmer friends struggling to make a living and about an eleven-year-old Cambodian girl I'd met whose farmer family had to move to the city, where I met her picking through trash in a dump. I told him about farmers who had lost loved ones to cancer they believed was linked to the agrochemicals they used in the fields. I told him that I was writing a book on how, if we got farming right, we could address these and many other issues, like climate change.

Freddy and Miguel began a long conversation. As Cliff and I sat in silence on the stone wall, the leathery leaves of the banana plants flapped in the breeze, sounding like waves meeting a shore.

Mamos are revered in Arhuaco society, representing nothing less than all living things as they work to maintain the balance of life on Earth. Each mamo has a gift. Some are prophets, foretelling the future in dreams. Others are healers. And some, like Freddy, are the equivalent of spiritual-agricultural extension agents, helping people live on and farm with the land by directing them on what to plant and when to plant it. They are chosen based on their lineages, or like Freddy, on their connection with nature. The mamos believe it's important to educate the outside world about how to live in balance with and as part of the natural world. That's where Miguel comes in.

## The Heart of the World

Finally, he turned to us. As Miguel thought, he would pinch his lips tight, dimples forming, before he would reference an American movie or political happening to help us understand what he was trying to explain. It was obvious that he was a big fan of *The Lord of the Rings* and saw its characters and themes—especially nature versus industrialization—as a useful way to explain the role of the Arhuaco to outsiders. He had spent more than a year in North Carolina and Washington, DC, and now lived forty miles away from this small village in the modern city of Valledupar, but he was rooted in the old ways. In a few months he would return to Pueblo Bello with his wife, who would deliver their first child under the guidance of the mamos. After the birth, they would hike up a peak to bury the placenta, sealing their child's connection with the mountains.

"The mamos are against books to teach you how to be," Miguel said, leaving me instantly deflated. The "against books" part didn't compute for me. Words, books, and stories are at the core of my life and career, and, after all, were why I was here in the first place. "You need to learn from life," he added. "More than write it down and write a book, change the way of living."

He continued, "We need to set an example. We need to start the change. A book alone won't inspire you to act. Change comes from knowing nature, life, the earth, and putting that into practice."

As Mark Shepard would put it, we needed to grow our AI (Actual Intelligence). And the way to do that was by experiencing and paying attention to the world around us.

"In the United States our relationship with nature is broken," I said. "I've returned because I know there is much more we can learn from you."

"If we give you the knowledge, are you going to take good care of it? Or are you going to use it for mining oil and gold?" Freddy asked.

The Arhuaco once lived closer to the sea and in fertile valleys, but many of them have been forced away from the abundance of the lower elevations and into the mountains. First there were the conquistadors, looking for gold. Then the missionaries, looking for souls. Companies came for oil, to frack, dam, and farm. They brought guns and disease, missionaries and mercenaries. And now we were here.

I understood why they might lump us in with such extractors, but each word of Freddy's blunt question landed like a punch. Cliff and I assured them we were there only to learn and share their message.

Finally, with some process and ceremony, Miguel handed us each a boll of cotton, his right hand to our right hands, and left to left. The unprocessed cotton was coarser and grayer than a cotton ball you'd buy at the store. The Arhuaco have grown cotton on the mountain slopes for generations. Cotton, as I would learn, represents their interwoven connection with their ancestors, among other things.

"We use cotton to record our good intentions," Miguel said. "It's like a book. Think of writing your goals, your fears, what you expect from life and nature in it. After that, Freddy will report to nature."

I closed my eyes and did my best to sit still. I'm a fidgeter, so it takes some effort, especially when I have no idea of how long a meditation will last. Most of the time when I meditate, I use an app on my phone. However, as the moments passed, the

coarse cotton in my palm felt heavier, or maybe I was imagining it. Once we'd filled the cotton with our thoughts, we were instructed to twirl it into the shape of a worm and blow on it. Freddy collected our intentions into his hands, blew on them himself, and talked under his breath to the mountains.

After twenty minutes, he addressed us.

"Nature and humans need each other," he said. "You came here, Kelsey. What you feel is nature calling you. And you answered that call. Our job as Arhuaco is to protect nature: animals, plants, and people."

Arhuaco spiritual leaders like Freddy (left) and Miguel are committed to educating visitors about how to live in balance with nature—but initially they weren't so sure about Kelsey and Cliff.
CLIFF RITCHEY

After more discussion, they said we could continue our journey if we worked with them, and simply didn't extract their images and stories for profit. They'd been down that road. We gratefully agreed to give them the final say in how they would appear in this book.

They tied cotton bracelets around our wrists. Miguel said they were like visas, opening doors to new places. We shouldn't remove them. When they fell off, we should tie them to a tree. And then Freddy shifted the way I saw myself and humanity in a few simple words, when he said: "The mountains are happy you are here."

•

The lesson began.

"It works like a satellite," Miguel added. "If mountains in the US are not protected, it is bad for all mountains. So our mission is to protect all of nature, everywhere. Each person, especially each Indigenous community, has a special job to care for a specific area.... We were given this area to protect."

I tried to process this, but got stuck in an infinite loop of thoughts and questions: "What specific area was I given to protect? Indiana? There are no mountains in Indiana." I've lived in four states, but I've felt the most connected to nature while hiking in the mountains or swimming in the ocean. When I think about protecting nature, I think about global issues, like the rainforest and carbon emissions, consuming less, decreasing my impact. Even if I knew where I was to protect, how was I to protect it?

Miguel gave us each another boll of cotton and Freddy provided our new instructions:

## The Heart of the World

"Now you are going to meditate about food and their connections. Bananas use water, soil, sun, moonlight. Let's think about cows. They need water, light, grass. Our body is full of life, life from the plants, life from the animals. We are consuming life."

He told us to weave our thoughts into the cotton.

I focused on bacon, as I often do. I thought about my neighbor's pastured pig out in the sunshine with a natural grin on its face eating grass, roots, and nuts. I thought about the grass. I thought about the pig's parents and about what they ate. Soon I was spiraling, considering the seeds that gave life to the grass, and the sun, and the water, and the microbes in the soil. It was overwhelming, like unraveling a spool of never-ending thread. Everything was connected, and it all came down to soil, sun, and water.

I'd never meant to close my eyes, but I had, and when I opened them I was staring at the earth. There are more living organisms in a teaspoon of healthy soil than there are people on the planet. There are bacteria, earthworms, beetles, ants, mites, fungi, yeasts, algae, protozoa, and nematodes. Mycorrhizal fungi, which live in the roots of plants, support a mutually advantageous network that transports water and nutrients, and pulls carbon into the ground. In fact, the soil stores more carbon than all plants and the atmosphere combined.

Arhuaco philosophies are clear that people are part of nature, part of this interconnectedness of all life. An interconnectedness that can't be undone because it's woven like a spiderweb. Pull on any thread and it impacts all the others and the web as a whole. In fact, maybe everything is so bound together that there are no parts, just the thing itself. We aren't just a part of nature. We *are* nature.

I saw the soil absorb the rain, the groundwater flow to a river. I saw myself from a river's perspective, and a heaviness and shame overcame me that I couldn't shake. I saw all the life that supported my own, and I felt a debt of gratitude that hardly could be repaid.

"Nature..." Freddy continued, "the life system is perfect. We are the imperfect ones. We have broken the rules of the Earth, and its perfect cycle. The Arhuaco are trying to be current with the natural cycles. The way we act fits with nature.... The problem is not the Earth. The Earth can live without us."

The breeze died; the banana leaves went silent. The air felt like a heavy blanket woven with guilt and responsibility, hopelessness and hope. We sat in the moment until Freddy decided we'd had enough lessons for the day.

•

The next day we headed into the mountains with Freddy and Miguel to learn from another mamo.

On my first visit, I'd had the privilege to visit Nabusimake, the heart of the Heart of the World, but Miguel informed me that it was now off-limits to us, even with our bracelets. Just as the Arhuaco had denied their business partners access, too many disrespectful tourists had forced them to change their policy despite the loss of income it represented.

"Nabusimake would be better," Miguel said, "but this will be very good. Every place has the magic."

Ahead of us, a donkey splashed through a mountain stream with our bags and supplies swaying on its back. I looked at Cliff, also

# The Heart of the World

loaded down with cameras, batteries, and solar chargers. His curly hair cascaded to his shoulders from his hat. He chatted with Miguel about a bird that had landed alongside the trail and about the cabin he built behind his house in Indiana where he could watch birds and sunsets. Miguel joked that Cliff, with his long hair and love of the natural world, might be an Arhuaco.

The donkey led us to a flat clearing near a multi-building homestead made of dirt and wood and in various stages of decay. It seemed that with every raindrop and gust of wind, the earth would reclaim a little bit more of it.

A family of five stopped their afternoon chores and stared at us. Miguel and Freddy approached an old man standing in front of one of the houses. My two companions reached into their satchels and placed several handfuls of dried coca leaves into the man's. He reciprocated with leaves of his own.

Every Arhuaco man carries a satchel of coca leaves and a hollowed gourd called a *poporo*, stuffed with crushed seashells. They insert a long stick, wet with their saliva, into the gourd to extract the crushed shells, which they rub on the coca leaves before stuffing them in their cheeks. The leaves of the mountains meet the shells of the sea, releasing a stimulant from the coca that helps with meditation and suppresses hunger. As they chew, they rub the stick around the rim of their *poporo*, each revolution recording a thought and conversation.

Arhuaco culture is based on producing, creating, and sharing gifts, not consuming. Coca is a sacred plant given to them by the creator. Sharing the leaves creates social cohesion and trust. This was only the first time on my journey that I noticed how Indigenous people saw plants and animals and features of the

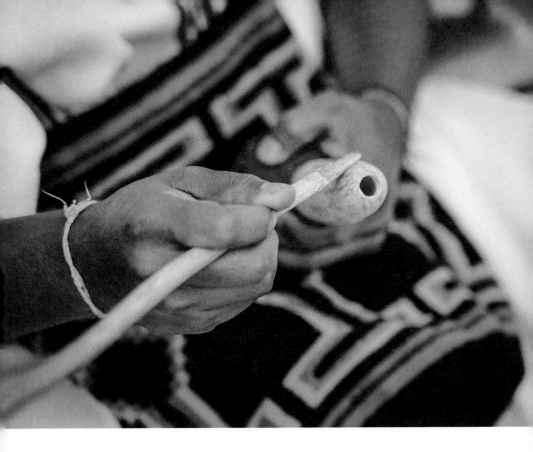

natural world as gifts, not something they own, but something to be grateful for and to share.

I could tell by the way Freddy and Miguel demurely shoveled coca into the old man's satchel that he must be Enemias, the mamo we were seeking.

There wasn't any sort of welcome or greeting directed at us. I don't think Enemias knew if we were worth knowing yet. His eyes were hazed with cataracts; his jaw was square and strong, but largely toothless. Single veins ran down each arm from his shoulders over lean biceps and tight forearms to old-man hands.

Miguel records his thoughts with each revolution of his stick around the rim of his *poporo*, a hollowed gourd. His handmade, one-of-a-kind satchel of coca leaves is on his lap. CLIFF RITCHEY

## The Heart of the World

Enemias nodded back to the trail and disappeared down the mountain.

It was as if he had jumped from the trail, grown wings, and flown away. Wings seemed necessary because the hillside was steep. Cliff and I followed Miguel, who pulled out a machete. I kept my weight on my heels so if I slipped and fell I would slide on my butt instead of tumbling and bowling them over. The ground leveled off and Miguel stepped through a curtain of reeds that closed behind him. We followed, entering a tiny clearing no more than ten feet wide. Miguel laid dead leaves on the ground for Cliff and me. Enemias was already sitting and spinning thoughts on his *poporo,* as if he had been there all day. He motioned for us to join him.

Freddy, Miguel, and Enemias sat cross-legged facing one another, their knees nearly touching and their voices barely audible over the noise of flowing water somewhere in the distance. Cliff and I sat just outside their circle.

I didn't suspect any human influence in this place. Without thinking, I had assumed it was wild mountain jungle, but as I looked around, I started to notice sugarcane, which blocked out the sun, stumps of previously harvested sugarcane, and banana plants with their red, purple, and pink flowers hanging down from spiky green tendrils like a plant in a Dr. Seuss book. We were surrounded by food.

"This seems like a wild place, but it's a place that they've shaped," I said, thinking out loud. This setting was different than our previous meditation spot. It was deeper into the forest. It felt closer to the heart.

"This is a sacred place," Enemias said, addressing us for the first time. Miguel translated and handed us each a hair from a coconut husk. "Take some time to think about your purpose in coming here. You may have some thoughts or questions to explore.

"Put your energy into it."

This was my third object, and first non-cotton one, to fill with my scattered thoughts, but even though by now I'd had some practice I still struggled to focus. I watched the three men whirl their sticks around the end of their *poporos*. There were peaceful and agitated rotations. They'd sync for a while and then one man would talk and whirl while the others slowed and listened. Their conversation was hushed by the sound of the unseen stream. The moist air tasted sweet with a hint of decomposition.

Eventually, my mind quieted as I sat there immersed in an abundance of life. I started to shrink and yet felt more connected. The longer I sat there, the smaller I became. Surrounded by reeds, sugarcane, and banana plants, I noticed highways of ants beneath the layers of decomposing leaves on the forest floor.

Before long, Enemias broke the silence. He wanted to know what we had thought about.

I had a story to share.

•

I had been at home "farming" America's number-one irrigated crop—grass—when I saw the snake in the tree.

My machine, my mower, hummed along on the remains of ancient life that was fed by the same sun I fought off with a wide-brimmed hat and a slick layer of sunscreen.

## The Heart of the World

There are more irrigated acres of grass farmed in the United States than corn, wheat, and fruit orchards combined. While tending to my roughly three acres of grass, I spill a little gas. Americans spill seventeen million gallons of gas each year filling up their mowers. On occasion I sprayed a little fertilizer, maybe a 'cide or two. Americans spread tens of millions of pounds of pesticides and fertilizers on their yards each year. Bright green, weedless yards of a single species of grass just might be the exact opposite of what regenerative agriculture is all about.

The snake was sunning itself on the branch of a tree. I posted a photo on social media, and went back to mowing, unaware of the stir it would cause. Turns out, people were offering a lot of advice: I could sell my house and move. I could burn down the tree with a flamethrower. My neighbor Heather grabbed her kids and locked down her house before texting me: "The snake needs to go!!!"

My wife, Annie, also texted me. She doesn't like snakes either. At one point, she pondered getting a snake gun, which is basically a handgun that shoots shotgun shells. She doesn't care much for guns, but she really doesn't like snakes. Why don't I just kill the snake? she suggested.

I didn't think a snake should be killed for being a snake.

This was during the early days of my regenerative journey, and I had just learned of a farmer in North Dakota named Gabe Brown. He talks about how he used to wake up thinking about what 'cide he'd have to use that day—managing life with death—but now he manages life with life. Since then, I'd begun to pause before reaching for Roundup or bug spray or a mouse trap. I had started to think that there was something wrong with the way our culture viewed and interacted with nature, and I could stand against it by not killing the snake.

Western myths, traditions, and beliefs are often rooted in the idea that nature is our dominion. A dominant story in my part of the Midwest involves a snake in a tree. "You may eat from any tree in the garden," God told Adam and Eve, "but do not eat of the tree of the knowledge of good and evil. If you do, you shall die."

"You will certainly not die," the snake in the tree later told them. "For God knows that when you eat of it your eyes will be opened, and *you will be like God*, knowing good and evil."

Of course, the snake wins the day, and we all got the boot. Temptation accepted; paradise lost.

Origin stories shape a culture, and smack-dab in God's flyover country in Indiana, the Garden of Eden was our origin story. In the beginning, God gave us the natural world to preside over and we learned quickly to fear a snake in a tree.

In contrast, many Indigenous origin stories involve humans, laterally positioned or subservient to all other creation, being granted life from nature.

Heather has a friend who works for the Department of Natural Resources (DNR). She sent him the photo of my mutant-death snake, and he was on the way.

I tried to think about snakes the way Gabe Brown, who manages life with life, would: Snakes eat moles. Snakes eat mice. When we kill one thing, it sends a ripple through the entire ecosystem.

Yep, I was willing to die on this hill, but I had to act fast. I grabbed a broom.

"Sorry, buddy," I said to the snake as I started to hit the branch. She wasn't getting the hint and wrapped around it tighter. "Come on! You don't want to go to animal jail." I got more aggressive and brushed her off the branch with the bristles. She dropped to the ground.

When the nature police rolled up, she was gone.

I was about to scream: "We live next to thirty acres of woods! There are snakes! They are an extremely important part of our ecosystem. They can live up to twenty years. It's mating season!"

I walked toward the DNR truck. I wanted to get this over with as quickly as possible.

"It was a black rat snake. I knocked it out of the tree and let it go on its way," I confessed.

"Looked like a python," he said.

"I'd really like to finish mowing," I said.

And then I used the one surefire way I knew to get him to leave. All I had to do was ask him one question: "But while you're here... how do I go about getting a beaver?" He stared at me as if he had never read a single book on the lumberjack rodent.

It's a long story, but I'd been working on this for months, and had even started some conversations about illegally smuggling a beaver across state borders.

The DNR guy told me most sane people ("sane" was implied) are trying to get rid of beavers. They chew down trees and flood ditches and streams and are pests that can be hard to get rid of. A beaver on the land was more trouble than a snake in a tree.

He left.

The next day, the snake came back.

•

Miguel finished translating my snake story. Silence. Enemias seemed lost in his thoughts, and resumed spinning stories around the end of his gourd.

Time passed. Enemias spoke, his voice as textured as a creaky door slowly opening. And then the breeze would blow, and then Freddy would say something, and more time would go by. It was hard to tell if they were having a conversation or talking to themselves, to each other, or to nature.

As peaceful and as mindful as I attempted to be, I grew impatient and whispered to Cliff, asking what he was thinking.

"When we sat down here and I was looking around noticing the ants and the bugs and the life all around us," Cliff said, "it reminded me of when I was a kid at my grandparents' house. There was a place like this with tall grasses. I would cut the grasses, and go be in the field all day long. Playing, being free, being with the ants and everything. It was a wonderful place."

I added that in our daily lives we are too busy and distracted to sit and notice.

"I wonder if we are in touch with nature when we are born, and then it gets baked out of us," I whispered.

We sat and thought about it.

After what felt like the length of a *Lord of the Rings* movie, Miguel spoke. "The mamo listened to your story. He says that snakes have their parents, too, that tell them what to do and what not to do. If they wanted, they would kill everybody. We need to respect their space. If we don't do anything to them, they won't bite us. People should understand that they are stronger than we are."

Admittedly, I was a little disappointed with the response. I was looking for some larger wisdom about our relationship with nature, but this was a little too snake-specific. And black rat snakes aren't venomous; maybe I should have mentioned that to Enemias. No one has ever been killed by a black rat snake. They certainly aren't going to rise up and kill everybody. Maybe something got lost in translation; maybe I put too much weight on every word Enemias said.

"Do you feel a connection with nature?" Freddy asked us.

Of course I did. Wasn't that the point of my story? I wasn't sure what to say without getting defensive. Thankfully, Cliff saved me by going into a beautiful tale about the significance of the great blue heron to his relationship with his wife, Julie. On their first date they saw a great blue heron. And they noticed one every time they were together. As they stood beneath a trellis at their wedding in their backyard, a pair flew over. Cliff pulled up his sleeve and showed Enemias his tattoo of the bird.

Enemias and Freddy mulled over Cliff's story, then had an oddly specific follow-up question: "Do heron eat fish?"

"Yes," Cliff said.

They processed this additional info. Minutes went by. I imagined them punching variables into some spiritual-environmental equation. And then they asked another follow-up question: "Do you have kids?"

"Yes," Cliff said.

This inspired discussion among them.

*What did the heron mean?* I was already jealous of the amazing answer I thought Cliff was about to receive. The Arhuaco spun thoughts on their gourds as our anticipation built. All at once they stopped and asked, "What do you think it means?"

Cliff shared that his first marriage had ended painfully. He hinted that Julie's first marriage had had a similar ending. When they started dating, "we were afraid at first," Cliff said. "But then we'd see a blue heron. We thought it meant courage."

Freddy gave a single nod of satisfaction as if to say, "You are welcome," and said nothing more.

Enemias stood. The afternoon sun lit him perfectly. I could see Cliff's camera finger twitching, but he hadn't been granted permission to take photos yet. Enemias, holding his gourd, was perfectly silhouetted, yet somehow there was still light reflecting on his face. He stood as still as a tree.

Once again, Miguel handed us cotton bolls.

"Our thoughts are like clothes," Enemias said. "They can be good or bad. White or black. Those thoughts fill our houses, where we work, how we live. Try to feel the bad thoughts that influence our lives. Sometimes we are just full of anger and we put that onto others. Put that bad energy into the cotton."

For thirty minutes I filled the cotton with my doubts, shortcomings, anger, but mostly my fear for the future of our world.

Enemias didn't ask us to share. He stepped aside.

"This is a sacred place," the mamo Enemias said, urging Kelsey and Cliff to take their time to think, and think...and think. CLIFF RITCHEY

"Toss the cotton in the water now," Miguel said. "Let it go."

We had been sitting there for hours and, although I had heard it, I hadn't realized the stream was less than ten feet away, running through a gully it had cut into the bare rock of the mountain. We walked fifty feet upstream, where a waterfall thundered into a pool not much larger than a hot tub. We tossed the cotton into the water and it disappeared downstream.

Miguel told us to get in the water, too. As I slipped in, Cliff encouraged caution. Water is powerful and you don't always know what lies beneath. My leg could get caught and I could be pulled under or I could be carried down the boulders to follow the piece of cotton.

The water, melted from the mountaintops above, was cold and my skin instantly numbed and reddened. I blindly felt beneath the foam and found a rocky seat, millions of years in the making. I sat and stretched my legs out to discover a place to press against, allowing me to hold my back in the waterfall. The water massaged my tired muscles. I fit in this space. It fit me, and as my tiny bit of cotton traveled toward the ocean, I felt lighter.

•

Bathing in the water of melted snow and glaciers, I couldn't help but think of a previous evening around a campfire with Miguel, Freddy, and a different, and even older mamo. We were in a small settlement, where some of Miguel's relatives lived, talking about climate change.

"People are not respectful of Mother Earth. Look at the rivers. Look at the species. Do you think it's normal for species to disappear or rivers to run dry? Humans believe we are more intelligent than nature. We are not. Everything that is happening is an answer to us thinking we are smarter than nature.

"The way we think is the way we act," Miguel said.

"People believe they can grow more food in a shorter time using pesticides or changing rivers," Freddy said. "That's the opposite of development."

But what to do about it? He continued, "The solution is not in the universities. We have the solution. It's in knowledge and culture. It's the way we act. It's our decisions."

The Arhuaco take time to consider how their actions will impact other living entities, the Heart of the World, and therefore the entire world. Miguel pointed to the tree above us from which a bird had pooped on me. The tree could be used to make furniture, build a house, or to burn as firewood. To cut it down would require a saw, fuel, transport, and human labor, but there are also externalized costs that our human economies rarely factor. The tree is also home to other living things, birds and insects that pollinate and spread seeds. The tree sequesters carbon and exhales oxygen. Some ecologists refer to these benefits as ecosystem services, and to remove the tree would come with additional costs that aren't directly paid for but externalized throughout the ecosystem. Nature provides humans with fresh air, water, food, and medicine. In 1997, *Nature* published a paper that attempted to put a tangible value on the intangible services of nature, and estimated that ecosystem services provided $33 trillion of value, which was nearly double the value the global economy produced. Miguel didn't cite the paper, but he did mention that the birds and insects bring messages to the mamo, and that his nieces and nephews love to climb trees.

It wasn't just our way of agriculture that needed to be changed. It was our culture, the way we think about nature. To emphasize the point, Miguel asked a simple yet unanswerable question that I'll never forget: "Why are churches sacred and rivers not?"

Our discussion continued and the old mamo seemed to be asleep.

An old mamo shares his wisdom that Earth will be okay; it will live on after humans and their destructive practices. As smoke drifts into his face, tears run down his wrinkled cheeks.  CLIFF RITCHEY

## The Heart of the World

"Are you hopeful?" I asked our circle.

As we talked, Miguel's nephew stomped a toy T-Rex through the dirt on the edge of the fire.

"When we are born, air... water is free," the old mamo started to speak before his eyes were even open. "Your society acts like you don't care about it. You are destroying things. Where are your kids going to live and eat if you destroy nature?"

The old mamo grabbed a stick and played in the fire, smoke drifting into his eyes. Miguel said that smoke no longer bothered the old mamo even as tears ran down the tributaries of wrinkles on his cheeks.

"I'm not worried. I'm acting properly," the old mamo said as if talking down to a naïve younger sibling. "I've been a good father. You are worried because you are here. Change yourself."

The old mamo picked up a drum and beat a rhythm, which he said was the rhythm of snowflakes landing on heaven. The little boy put down the T-Rex and clapped his hands.

•

"The moon," a voice whispered in the night.

Someone shined a light into my tent pitched among the earthen buildings that made up Enemias's homestead. I blinked and squinted. A full moon perched in the mountains to the east and lit everything like a spotlight. I crawled out of the tent.

"Good morning, Kelsey," Miguel whispered, sitting on a log next to Enemias.

The fire flickered on their faces, but there was no darkness behind them. Everything was silver with moonlight. I turned my back to the moon and followed my shadow to a seat by the fire.

Bathing in the moonlight, by the time I sat I was as awake as if I had jumped into the cold stream again. The insects and frogs sang a song similar to those in Indiana but with a different accent.

Enemias looked at me and gestured wildly as if he were communing with invisible forces. He talked louder and more than he had all day.

I hoped he would offer a more satisfactory interpretation of the snake in the tree, but whatever he was saying, Miguel wasn't translating it.

Suddenly I had a thought. I put in an earbud and found the recording I had of him telling me about the snake. Listening again, I realized I had missed his final words. After warning me that snakes can hurt us and we should respect them, he'd continued:

"A snake can bite a lot of humans," he'd said. "And we cannot fight them. We don't have that power.... If we are against nature, it is going to act against us."

Oceans rise, wildfires blaze, record high temperatures are broken. Nature is already acting against us. I carried the anxiety of our accumulating crises into these mountains. I still felt their weight, but in the moonlight a new thought emerged: If we act with nature, it is going to act with us.

The message of the Arhuaco is that the Earth can and will live without us, but we cannot live without it. And that requires belonging to nature, having their and Mark Shepard's Actual

**The Heart of the World**

Intelligence, and embracing experiences and a relationship with the natural world so that, within those experiences, we can find answers to our questions.

We may die off, but the heart of the world will keep beating.

# Part I: Reflect

# Chapter 3
# Root and Plow

# Corn Belt, USA

**Regenerative Agriculture Requires Long-Term Thinking**

I rolled a Smurf-colored soybean seed in the palm of my hand. I was too distracted by its hue to think about the sun, soil, and water—the entire universe of relationships that led to its existence. All I wanted to know was: Why was it blue?

"You shouldn't touch those. They're coated," my friend said.

I looked up at him and hurriedly dropped the seed to the ground like it was burning my skin.

"You'd better wash your hands," he added.

The seeds were covered in poison, likely a fungicide that would protect them from the wet soil of an early-spring planting. I knew there were plenty of harmful chemicals employed in conventional agriculture, but I hadn't dreamed that the seeds themselves would be poisonous. Farmers are advised to avoid inhaling any dust and to wear long sleeves, pants, and rubber gloves while handling them.

Planting uncoated corn and soybean seeds is considered "planting naked," exposing them, your yield, and your livelihood to the risks posed by nature. The EPA doesn't regulate these blue chemicals as much as the ones used in in-field applications, and therefore the amounts used and environmental impact aren't well known or tracked. What's more, their actual impact on yield is somewhat dubious.

These widely used blue-colored seeds aren't Smurf eggs, but soybean seeds coated in poisonous fungicide. The environmental impact of the fungicide is unknown. LYROKY / ALAMY

Still, the agrochemical companies are well aware that adding "value" to their product, and charging $10 to $15 more per acre of seed, positively impacts their bottom line. Bayer, Syngenta, Corteva, and other companies provide replant insurance: If you use Smurf seed and your crop fails due to weather or pest, they'll cover 100 percent of the cost of the seed to replant. However, if you are "planting naked," the coverage can drop to half that.

My roots run deep here in western Ohio near the Indiana border. From where I stood, I could see my parents' first house and the cemetery where my wife's grandmother was buried. Just beyond that was the high school where I was on the basketball team, my childhood home, and my entire world until the age of eighteen. I was here to learn about and experience the most common type of farming in the United States. And I was here to reconnect with a friend.

As a kid I used to sit on my grandpa's simple tractor and pretend it was a spaceship, even though it only had a steering wheel, throttle, brake, clutch, and gearshift. But my friend's tractor—well, it was a super spaceship, cool to me even now. It had screens that monitored the rate the seeds were dropping from the planter, and computers set the speed to maximize fuel efficiency. He didn't even have to steer. GPS did that. The company he bought the seeds from had a specific plan for each of his fields, based on data collected over the past eight years.

He didn't want to go on the record with me. In fact, he'd even been reluctant to see me at all, which didn't feel good. We had been on the same basketball team. His dad used to call and chat with my mom on the phone when they were kids.

He wasn't alone. All the farmers I'd gone to high school with were less than enthusiastic about talking to me at first, and none would let me use their name. I worried that they viewed me as someone who would judge them personally based on their practices and profession, seeing me as an over-read idealist who didn't know a thing about real farming. Or maybe they saw me as a threat that could cost them their job, clients, and standing in their community. They had no idea that I was proud of them for staying and building the community I had fled.

All of them, at least second-generation farmers, were involved in running family farms of several thousand acres and will make decisions that will shape the future of farming and their families. They control the acreage, so if there's going to be a substantial shift to regenerative agriculture, they'll need to be part of it.

"What we're doing now, this is the reward," my friend said. The agrochemical company had written the prescription for the land—specifying the seed spacing, the amount and placement of chemical applications—but his job still required a lot of planning. He rented this land, so he had to work with the landowner on a lease, order all his inputs, make sure his equipment was running, and find a good time to plant. He had to juggle a lot of unknowns before he could get to what most kids in our area thought was the coolest part of farming: driving a tractor.

"Farming is a decision-making process. This is pressing buttons," he continued, pointing to his spaceship. "We have to make all the steps work together, make sure the landowners are happy, make sure that we are doing the best we can for Mother Nature and the ground."

This needs saying: My friend and the other farmers who are still farming are highly skilled at what they do. Somehow, against the odds, they and their families have navigated an industry in which they have to farm more to earn less and deal with a changing climate and the extreme weather events that come with it. They also had to weather the impacts of the farm crisis of the 1980s, in which plummeting land values and soaring interest rates pushed 10 percent of farmers off their land, and who knows how many family issues, dramas, and tragedies. They are doing the best they can in a system that is far worse than I had imagined. Industrial agriculture accounts for 31 percent of global greenhouse gas emissions, flushes topsoil down our rivers at an alarming and unsustainable rate, feeds nitrogen to the dead zone in the Gulf of Mexico that some years grows to the size of New Jersey, and squeezes small farmers off the land.

The agrochemical companies have yield contests pitting farmers against one another. It's a source of pride at the coffee shop to have the highest. Yet, the higher your yield, and the higher your neighbor's yield, the lower the price you'll get. The cost of the seed—new and improved, now with higher yields(!)—goes up. The cost of the pesticides and herbicides with similar promises go up. And if yields are high and prices are low, you must have even higher yields if you're going to stay afloat, so you spend even more on the latest and greatest seed and chemicals.

Mark Shepard had shown me that a different type of agriculture was possible in the hills of Wisconsin, but I wondered if his vision was transferrable to the flatlands of the Corn Belt. And I was curious whether these farmers in the Midwest so enmeshed in the Big Ag system had concerns about the way of farming they had inherited.

## Root and Plow

It was time for me to have the talks I'd been avoiding for years. I was fearful that my concerns about the way my old friends, family, and neighbors farmed might not be received well and could break the remaining ties I had to the community where my parents and in-laws live, the community that raised me.

To my relief, the conversations I had in the cabs of tractors, in turkey barns, standing in fields, and sitting in living rooms went well. Once the farmers knew I was genuinely interested in their perspective and not trying to bash their profession, they revealed that they too were questioning our entire agriculture system. They loved farming, but weren't sure how much longer it would make sense.

What I decided to do, to include their voices but protect their identities, is to throw our conversations into a blender and create a general voice expressing their view, which I'll call Ohio Farmer. Note: The following are real quotes from real people.

We started with an economics lesson.

> **Ohio Farmer:** *My grandpa farmed by horse, but today farming has become big business. We've had to grow in order to keep the farm. It takes me thousands of acres to support my family. And we've just been breaking even the last few years. I haven't taken an income from the farm in the last eight years. Many farmers or their spouses have off-farm jobs.*
>
> **Kelsey:** *What's a good net income per acre?*
>
> **Ohio Farmer:** *I usually budget to make around $250 per acre.*

**Kelsey:** *You farm a whole acre and only make $250 for the year? This year I planted fifty pumpkin plants on a tenth of an acre. Each plant can yield five pumpkins that I could sell for $5, which comes out to $1,250 from one-tenth of the land.*

**Ohio Farmer:** *But how many pumpkins did you actually grow?*

**Kelsey:** *Well... The deer ate the flowers, or we didn't water them enough, or I used bad seed. I'm not sure what happened. I was in Brazil most of the summer.*

**Ohio Farmer:** *So how many pumpkins did you get?*

**Kelsey:** *Two.*

**Ohio Farmer:** *And how much did you sell them for?*

**Kelsey:** *I didn't sell them. I just put them on our front porch and our dog Jersey ate them after a day.*

**Ohio Farmer:** *So one-tenth of an acre yielded $0.*

**Kelsey:** *Farming is hard.*

It got worse.

**Ohio Farmer:** *Exactly. Today farming is an oligopoly.*

**Kelsey:** *I get my "opolies" mixed up....*

**Ohio Farmer:** *Basically, there are about two companies that I can buy seed and inputs from. They can jack up the price and I have to deal with it. They barely have any competition, so the prices continue to rise. On the other end, when I go to sell my product,*

> *I'm competing against every farmer out there. The more we produce, the lower our prices go.*

And worse.

> **Kelsey:** *Do you ever eat your corn and soybeans?*
>
> **Ohio Farmer:** *No. But I eat the animals that eat it.*
>
> **Kelsey:** *What about the animals you raise? Do you eat them?*
>
> **Ohio Farmer:** *No. I don't like their meat.*

Many Ohio Farmers own a few Concentrated Animal Feeding Operations (CAFOs) and raise one type of animal for one stage of its life. Essentially, they own the building and get paid to manage it and the animals.

I wonder what keeps these folks going.

> **Ohio Farmer:** *We have a growing population. We have to feed the world.*
>
> **Kelsey:** *What if I told you that half the world's calories are produced by small farmers?*
>
> **Ohio Farmer:** *I wasn't aware of that.*

Farmers who work less than five acres produce 35 percent of the world's food, despite cultivating only 12 percent of the land. A report in *Nature* reviewed 118 studies from fifty-one countries comparing small farms to industrial farms and found that small farms have higher yields and more diversity. The United Nations predicts that under our current food system, global farmers will need to increase production by 60 percent over the next three decades in order to feed a population of ten billion people. The thing is, our current system produces enough food to feed ten billion people already, but it wastes around one-third of that food, and some of it, like 40 percent of corn, is turned into biofuels such as ethanol. And still two billion people are food insecure. Our system is neither efficient nor just. The system is geared toward feeding profit, not people.

**Ohio Farmer:** *But do you eat cornflakes for breakfast?*

**Kelsey:** *Sometimes. I just bought some organic, hippie cornflakes.*

**Ohio Farmer:** *Corn and soy are also used in so many fabrics, foods, and plastics.*

**Kelsey:** *Maybe that's just because we grow so much of them. If I told my farm friends in Kenya or Colombia that farmers in Ohio don't eat the crops they raise, they'd look at me like I had two heads. Would you even be allowed to eat the livestock you raise if you wanted to? Some company owns them, correct?*

**Ohio Farmer:** *That's right. But they'd probably be okay if I asked them to have one or two.*

**Kelsey:** *Isn't it nuts that farmers don't even own the livestock they raise?*

**Ohio Farmer:** *It is, sort of.*

**Kelsey:** *I'm not sure our views are that different.*

But here's what blew my mind.

**Ohio Farmer:** *Honestly, I agree. We want people to prosper and have access to affordable and healthy food.*

*I do my best to treat my land with respect and not to pollute with chemicals. Chemicals are expensive. We use them a lot more efficiently than we used to. We have ground that we used to farm when I was a kid, but now we can no longer farm.*

**Kelsey:** *I don't doubt that you try your best to treat the land and people with respect. I don't, however, have a lot of faith in the agrochemical companies.*

**Ohio Farmer:** *You know... Dad was doused with Roundup when a hose blew. Grandpa thinks that's what led to the cancer that killed him.*

Signs advertise an agrochemical seed brand to passing traffic. Field corn like this won't appear in the grocery store. In fact, 40 percent of it is destined to be burned by cars as ethanol.
JIM WEST / ALAMY

I went to his dad's funeral. It wasn't the first funeral of a farmer where whispers like this were passed around.

> **Ohio Farmer:** *The companies tell us the chemistry is safe and that there's never been a connection between cancer and the chemicals we spray. When there's a lawsuit, companies settle because it's easier than fighting it in court.*
>
> **Kelsey:** *Let's say you saved your own seed. Let's say you planted a diverse set of crops so you'd have less pest pressure and needed to buy less chemicals. Let's say your yield was reduced some, but your cost of inputs went down and your profit went up. You make more money and agrochemical companies make less.*
>
> **Ohio Farmer:** *I think most farmers are stewards of the land. A regenerative farm is the ideal. I'm not against it. But we don't live in an ideal world. I farm how I know how to farm. Also, we can't save our seeds. It's illegal.*
>
> **Kelsey:** *I've heard that some farmers are concerned that soon the corporations won't just own the seed, but corporations will fully own and manage the land, too. Do you think that's a possibility?*
>
> **Ohio Farmer:** *Yes. There are companies that contract the land and the people who farm it are their employees. So people or corporations buy land as an investment and then just hire a land-management company to deal with the rest.*
>
> **Kelsey:** *So how do you compete with that? You'll have to continue to grow?*

> **Ohio Farmer:** *Yes. In the future there are going to be hobby farmers who farm three hundred to four hundred acres and then there are going to be farmers who farm five to ten thousand acres. And of course there are going to be large factory farms that own and manage tens of thousands of acres. There's going to be a hollowing of the mid-size farms of one to two thousand acres.*

The future—including a "hobby" farm on more than $5 million of land—didn't sound so great to me.

> **Kelsey:** *Farming involves a lot of hard work and uncertainty for very little reward at times. Do you think people still want to farm, if given the opportunity, or are we destined to be fed by corporate farms?*
>
> **Ohio Farmer:** *That's a good question. Find out and let me know.*

And off I went to Kansas and Minnesota to talk to farmers who were actively working toward a different future of agriculture, and these farmers looked like, talked like, and shook hands like the farmers in my community. They were grain farmers, questioning twelve thousand years of agriculture and growing an unlikely crop, a perennial grain known as Kernza®.

•

### Salina, Kansas

Often our career choices are rooted in earlier challenges. A sick kid becomes a doctor; an injured athlete, a physical therapist. Lee DeHaan, the son of farmers who lost their farm in the '80s, grew up to be a crop scientist.

A shirt that read "Root Down" hung off Lee's prominent collarbones. He stood in a high-tech greenhouse surrounded by gently rolling farmland in Salina, Kansas. Fans hummed. There were grow lights, battery backups, and emergency lighting. At night, the building glows red. Wispy blades of grass grew from plant trays sitting on knee-high tables.

Farmers like my grandpa raised a variety of animals and crops, supporting themselves on a hundred to a few hundred acres, but then everything changed for American farmers in 1973, when Secretary of Agriculture Earl Butz told farmers to "get big, or get out." They should use every bit of land and plant "fencerow to fencerow." Butz, who undid previous government controls to avoid overproduction and price collapse, wanted an abundance of commodity corn, soybeans, and wheat to feed a growing middle class with cheap food. And of course, this would require large quantities of agrochemical inputs and benefit the large food conglomerates, several of which Butz served as a board member. Butz aligned the national interest with the interests of Big Ag.

A whole generation of farmers, like Lee's dad, didn't want to get out, so they tried to get big. More land, more equipment. More debt. Some farmers found they had to ask banks for more than they needed in order to get a loan. Even more debt. And then the increase in acreage and yields led to overproduction driving prices down by the early 1980s. Speculators had been driving up the prices of farmland, but because of the grain surplus, land values collapsed up to 60 percent from 1981 to 1985. Interest rates skyrocketed to 21 percent, which crushed farmers. The number of farmers in the United States dropped from 3.7 million to 2.2 million. The deck was stacked against Lee's dad.

## Root and Plow

Lee remembered the day his older brother brought home John Mellencamp's 1985 album *Scarecrow*, featuring "Rain on the Scarecrow," an elegy to the farmers who faced the 1980s farm crisis. In the song a father remembers his own grandfather, who had cleared the family's land. He expresses his pride in being able to feed a nation. But when crop prices fell, he no longer made enough to pay the loans and buy seed, and the bank foreclosed. He apologizes to his son for losing their land and legacy, for leaving him nothing more than memories.

Lee's dad was on the same path. But he was able to sell his Minnesota farm to an investor; he no longer owned the ground he farmed.

Then in the early '90s he heard a visionary speaker describe a new future for farming—perennial agriculture—and how it could improve the quality of the land and cut costs by using less fertilizer, less chemicals, less plowing. When people thought of perennial agriculture they thought of orchards and smaller vegetable operations, but a farm of any size growing perennial grains was unheard of.

The speaker saw annual agriculture as a crisis. He quoted the Bible: "and they shall beat their swords into plowshares, and their spears into pruning hooks: nation shall not lift up sword against nation, neither shall they learn war anymore."

And then he added something like, "The plow has destroyed more options for future generations than the sword. For all the destructiveness of war, the loss of soil and soil fertility is potentially more devastating."

Lee grew up hearing his father and one of his older brothers talking about these ideas. Robb made a big impression on him

The Land Institute's Lee DeHaan is working to develop the commercial viability of Kernza, a perennial grain. Unlike annual grains, perennials don't require planting each year, reducing labor, fuel, and soil disturbance. AMY KUMLER

## Root and Plow

when he said simply that the world needed "different kinds of plants." Robb studied perennial plants at the University of Minnesota, but struggled to get funding, since many of the grants supported annual agriculture. When it was time for Lee to start his PhD there, he was able to take advantage of their new Forever Green Initiative, which focused on developing and improving winter-hardy annual and perennial crops. He was doing research on seeds his brother had left for him.

"The idea of the Forever Green Initiative is that you have to keep your land green at all times," Lee said while I ran through my developing mental checklist of core regenerative ideals. At The Land Institute, the idea of "farming naked" is also frowned upon, but to them it doesn't mean planting an uncoated seed, it's leaving your land bare. "If it's brown," he said, "you're wasting sunlight, you're wasting water, you're losing soil—"

"Um," I said. "Someone's knocking at the door."

Lee couldn't hear it over the fans.

"You must not have the new password?" Lee said as he opened the door.

"They wouldn't give it to a guy like me," the man said in a low voice graveled with experience. He walked in, grabbed a Shop-Vac, and wasn't pleased about the water still in it.

I introduced myself.

"Wes Jackson," the man said. Wes was the speaker who Lee's dad and brother had listened to in the '90s, the man largely responsible for the path of Lee's career and the founding of The Land Institute, where Lee now worked as Director of Crop Improvement.

Wes is part of Lee's story and Lee is part of Wes's.

The Land Institute is a nonprofit research organization working to develop farming systems that use diverse crops and mimic nature, especially the prairies that dominated the Midwestern landscape before the domination of corn and soybeans. Through The Land Institute, Wes and his team have worked for decades toward developing perennial grains as commercially viable crops. In an essay in his book *Nature as Measure*, Wes explains that "humans had long gathered and eaten [the] seed of many herbaceous perennial species, especially grasses, but the domestication step that could have generated perennial grain crops never happened...." Instead, twelve thousand years ago, when agriculture began, early farmers chose to focus on annuals, which made it easier for them to select plants with attributes they desired, such as the ability to hold onto their seed after it ripens, leading to an easier harvest.

Lee explained it to me like this. Prairies were naturally sustainable. They were home to a diversity of edible perennials and annuals, and with the help of animals like Mark Shepard's bison and mastodons, created soil. An array of plants thrived. Things were working fine for thousands of years, and then something screwed it all up. That thing was us. When humans discovered that we could cultivate annuals like corn and wheat, some gave up hunting and gathering, which takes a lot of energy, and settled into one place. In this way, they could control their food supply, even growing a surplus. This shift was our original agricultural sin.

That new approach led to what we do now: We feed a significant amount of corn and other plants that humans can eat to animals.

Then we eat the animals, losing up to 90 percent of the food's energy in the process. Yet some of those animals are perfectly able to thrive on plants that humans cannot digest and are located on terrain where our food can't be grown. We likely couldn't have designed a more inefficient system if we'd tried. Lee refers to it as "absolutely stupid" and "total nonsense."

Wes makes the point that most people focus on the problems in agriculture while he and The Land Institute address the problem *of* agriculture. The shift to annual monocultures of grain required open areas for planting and growing, which meant extensive clearing of the land. Hoes became sixty-foot plows and human labor was replaced by beasts of burden and later by tractors, monocultures grew in size, and more soil and the life that lived in it was lost. While annual agriculture and domestication created food reserves that led to modern civilization, they also threaten it. As the scale of agriculture has expanded to feed the planet's growing population, that threat has only increased.

While The Land Institute believes that from the very beginning, agriculture has degenerated the land, their mission imagines a future that incorporates perennials and minimizes tillage to "build soil and people rather than degrade and deplete them," rely on sunlight more than fossil fuels, "foster biodiversity from microbes to megafauna," and "store carbon in the ground while more resilient human cultures grow from the ground up."

To me, that sounded like an agriculture that's regenerative. And one that was outcome-based, which was different from the narrow, process-based definitions I'd seen from the agrochemical companies. They act like adopting a single process—such as cover cropping or practicing no-till, regardless of the amount of agrochemicals used or the utter lack of crop diversity—makes for a regenerative agriculture. The corporations throw the term around when of course they continue to sell the same old degenerative products. I had to wonder if anyone cares about the process if it doesn't rebuild the soil and the life that depends on it.

Lee told me that most of what has happened in agriculture since World War II can't be explained by biology, but by politics and economics.

For example, he said, "We started subsidizing grain production and inputs, and reducing the cost of oil. And pretty soon we're pumping oil in western Kansas to grow corn to make ethanol to burn in our gas tanks. That results in less energy in the end than if we had never pumped out the oil to begin with. We're using oil, pumping carbon into the atmosphere, causing global warming. Nothing about it makes sense."

While twelve thousand years ago early farmers chose annuals over perennials, Wes, Lee, and The Land Institute use modern science and technology to go the other way. They are working to domesticate perennial grains, such as the wispy green grass on the tables in the greenhouse, to address the original problem of agriculture.

Wes dumped out the water from the Shop-Vac down a drain and hurried out the door. It was a busy weekend at The Land Institute. It was Prairie Festival, when people from all over the world pile into a drafty barn outside Salina to hear scientists, authors, experts, and artists talk about the future of agriculture.

Lee was recruited to The Land Institute after his PhD program. It was one of the few places that studied perennial grains. Most

Wes Jackson, a founder of The Land Institute, showing off the long roots of Kernza, which pump carbon deep into the soil while helping to retain topsoil and water. JIM RICHARDSON

research jobs in agriculture, especially the ones with agrochemical companies, focus on plants and chemicals they can sell to the same farmer year after year. But Lee thought there might be money and benefits for small farmers like his dad, who "was never an alternative farmer," Lee said. "But he always wanted to use the most sustainable technique that he could." Just like my friends in Ohio.

But, Lee continues, "When it comes to our work on perennials, the reaction I usually get is something like, 'If you can do that, if that works, that'd be amazing.' Farmers fundamentally see the beauty of it."

"Pardon my ignorance," I said, pointing to the grassy plants on the low tables. "Is this Kernza?"

"To be really careful with my language," Lee said, "this is wheatgrass…. Grain from this species has been sold under the trademark name of Kernza."

Basically, the plant is a perennial cousin to annual wheat and is often grown as fodder for livestock. In 1983, after hearing Wes's vision of a perennial grain and before The Land Institute itself had such capabilities, plant breeders at the Rodale Institute started working with the forage grass. In 2003, Lee and The Land Institute took over the efforts.

Once they had a viable product that could be used in foods, Lee and his co-researcher David Van Tassel came up with the name Kernza as a nod to the Konza Prairie, an 8,600-acre conservancy of tall-grass prairie near Manhattan, Kansas. Lee also thought Kernza sounded like a grain. He and his team are still focused on increasing the size of the grain and the overall yield. The trademark allows them to keep a handle on how Kernza is being raised and to build a market for it.

A plant's roots pump carbon down to feed the microorganisms that live in the soil, hold the soil, and hold water. More living roots in the ground means more carbon is pumped out of the air, and the soil becomes more nutrient-dense. Kernza's long roots—four times longer than the part of the plant that's above the surface—are alive and contributing to soil health year-round. Annuals like corn and soybeans—with much shorter roots—have living roots in the ground for only about four months.

Kernza's deep roots can also help protect wells from the dangers of nitrate leaching—caused by excess chemical fertilizers and manure in fields of soy and corn—in places like Minnesota, where 49 percent of wells in agricultural areas have higher concentrations of nitrates than the EPA standard. Nitrates in drinking water have been linked to cancer and blue-baby syndrome. One study showed that in only three years, Kernza planted near wells in a former cornfield cut nitrate contamination by 96 percent and by 86 percent in a former soybean field.

On top of that, Kernza has more protein, fiber, bran, vitamins, and antioxidants than wheat. Its sweet and nutty flavor can be found in breakfast cereals, beers, pasta, and breads. One of its first commercial uses was in a trial run of tortillas for Chipotle.

But is it ready for the big leagues?

Later that weekend, during Prairie Festival, I was part of a group that rotated from one in-the-field presentation to the next. Lee stood holding a mic at the back of a red pickup truck with a small speaker on its tailgate. He held up a bag of Kernza seed and rattled off facts. At the end he took questions.

"So if you got a field of standard Kansas wheat, how many bushels would you have gotten per acre?" a man with a white beard asked.

"Kansas wheat this year, you'd get around forty-five bushels per acre."

"How many bushels of Kernza would you average?" the man asked.

"My best stuff would be about twenty-five percent of that," Lee said.

"So, it's nowhere near production." The man verbalized the argument I've heard others level. Kernza's got a lot going for it, but what it lacks is yield.

Lee takes the long view. He recognizes that his work isn't just about *now*. Though things are moving in the right direction, he told the man it's going to be years before Kernza will have the yield of good ol' Kansas wheat.

I watched the man's face fall as he registered Lee's response. He realized he'd be long dead before he could be driving through acres of amber waves of Kansas Kernza grain.

He might have a different reaction if he visited the farm of one of Kernza's earliest farmers and champions, where yield is not the only topic.

•

### A-Frame Farm (Madison, Minnesota)

After lunch on his back porch in western Minnesota, Carmen Fernholz and I walked into in his barn, where a young farmer named Luke poked his head into the guts of a 1988 John Deere combine.

"We've come to watch you work," Carmen said.

"I might need some advice," Luke said, pausing to look around the side of the machine.

"I was thinking the same," Carmen said.

Carmen bought this eighty-acre farm back in 1972 for $25,000. It was sort of a blank slate. Over the years he built a house, a barn, and planted trees next to the long drive as he developed A-Frame Farm, where he and his wife, Sally, grew certified organic corn, soybeans, and small grains—tiny cereal crops such as oats, barley, and alfalfa.

Now he was in his seventies and his kids had no interest in taking the farm over. He didn't want to sell to someone who didn't share his values and land ethic. So, Luke was an answered prayer.

"As an old-timer, you always want to be optimistic," Carmen said, "and I am. Especially when you work alongside people like Luke, who are interested in the connection between the soil and the person eating and everything that takes place in between. In 1975, when I was first doing organic, you didn't even say the 'O-word.' And today it's front-page. We know people can change. We know that the system can change. We have to have patience. And we have to keep pushing."

Carmen's farmer father had wanted his children to take another path. Farming meant too many highs and lows. Having only gone through eighth grade himself, he valued education and he never stopped reading and learning. He had copies of Rodale's organic farming and gardening magazine around the house.

When Carmen transitioned to organic farming, he didn't use the O-word. He'd edit himself to keep the conversation going; he'd talk about the benefits of planting cover crops, such as rye, buckwheat, or alfalfa in between cash-crop growing seasons to improve soil health. He was known as the "crazy farmer east of Madison," but found encouragement through a network of other crazy farmers, and a cheerleader in the local poet Robert Bly, who wrote about rural life in western Minnesota.

While he stepped onto the agrochemical treadmill, like most of his peers, Carmen's dad flagged off an acre of wheat for the family's consumption and used no chemicals there or in the family garden. As a kid, Carmen would overhear conversations about "organic," and the fact that they didn't use chemicals on the food they ate stuck with him.

Carmen went to a liberal arts college, which opened his mind. But after graduation, he was floundering. His job teaching English, speech, and reporting wasn't working out; he dabbled in real estate. He decided to go to grad school—and on the day he was supposed to start, he bought this eighty-acre farm instead. He says he felt like a different person when he walked on the land for the first time.

It was a stretch, but Carmen and Sally made it work. He picked up a part-time job as a carpenter for $1.75 per hour and she worked part-time in healthcare. They were focused. The first year, he borrowed his dad's equipment and farmed conventionally, using the chemical inputs that were growing in popularity because they showed such clear initial benefits and their long-term consequences weren't well known.

Then Carmen heard some old-timers saying they were not going to bend to the pressure to transition to chemicals. No one hoes this row alone, so they became his mentors as he turned away from conventional agriculture.

I think it's a shame that we call farming with genetically modified seeds and agrochemicals "conventional agriculture," as if it's normal to grow food with poisons developed for warfare. Humans had been farming without chemicals for twelve thousand years. Organic agriculture was the norm, the actual convention. But when chemical inputs took over, there was a need for non-adopters to differentiate themselves, so farmers like Carmen began calling their methods "organic," a term he'd been reading about in his dad's Rodale magazine.

"I'm a farmer," Carmen said, "and I emphasize that I'm a farmer. Not a majority, but a significant number of farmers are not environmentalists. If we were, we would strive to diversify the landscape. And I don't think we would see a lot of the farming practices that are actually going on."

Carmen worked with the Organic Growers and Buyers Association to develop early organic standards, hosting a sort of continental congress on the subject the same day a local agriculture agent threatened to kill one of his fields because there were too many weeds. Organic agriculture isn't for the faint of heart.

But Carmen made a life and a living at it. He became the mentor for so many, and eventually Luke.

Carmen wanted to say something to Luke, tell him what he thought needed to be done on the combine. He fought the urge to help several times. This was Luke's first year running the operation on his own, and after more than four decades of being in charge of what happened on his land and in his barn, Carmen was still adjusting.

Carmen had known of Luke for years. He had directed Luke's then-girlfriend and now-wife in a local production of *To Kill a Mockingbird*. The young couple had married right after high school. Luke went to Alaska and made chainsaw carvings. He moved back and worked for the Department of Natural Resources and later the local co-op, where he sold seed and agrochemicals for DeKalb, Syngenta, and Monsanto. Luke is soft-spoken, a gentleman who seems too gentle to be a salesman of any sort.

"It wasn't my type of business," Luke said. "I'm not very aggressive... in the sense of trying to undercut the next salesman or lying to buyers."

At the same time Luke was borrowing money to try to start farming himself, he noticed that the credit flowed more easily to farmers who stayed inside the lines. That meant corn and soybeans, so this was his initial path.

Luke's wife was a nurse practitioner who focused on preventative care. She started to focus on how food was produced and introduced Luke to a doctor who talked a lot about food as medicine and connecting his patients to locally sourced food.

Luke asked how he could help. The doctor looked at him and said, "You can start by growing real food."

That sat Luke back in his chair. So he called Carmen, who lived a few miles away, and asked if organic farming was possible. Of course it was. Carmen had been doing it since 1973. Luke asked if he could work alongside him for a year, unpaid, to learn.

Luke (left) and his mentor, Carmen, in Madison, Minnesota. Carmen's children didn't want to take over the farm, so when Luke reached out wanting to know more about organic farming, a partnership was born. COURTESY OF DODD DEMAS

Carmen agreed, and at the end of the year they created a plan where Luke would start to buy out Carmen's equipment and rent his land.

When Carmen and Luke talk about their plans, both of them say "we." "We did this.... We'll do that."

Luke finished the combine repair, and the two men talked about gates and cylinders and chains for a while before Luke climbed into the cab, and it was time for Carmen to show me his forty acres of Kernza.

We drove past the home he'd grown up in, where his sisters, both nuns, now live at and operate Earthrise Farm. The Earthrise sign, shaped like the sun, bore a picture of a plant with the Earth as the bloom along with this quote from one of the first astronauts on the moon: "We have seen the splendor of Earth rise above the horizon of the moon."

Carmen parked in a ditch next to the Kernza, which didn't look much different to me than a field of annual wheat. It's just taller and wavier. He got out and drew a map of his land on the dust of his truck's bumper, going into great detail about how each location was different. He told me I could pick out any five acres and he could describe it exactly.

"It really comes down to bonding with the land. That's what drove me to eliminate fall tillage," Carmen said. Tilling in the fall exposes loose soil to months of wind and erosion. "I began to understand more and more, what's going on in the soil. And then to watch the interaction, the relationship between the weather and the crop, and then the seasons, how they come and go.... You get to know every inch of the land."

**Root and Plow**

With regenerative and organic farming, you have to pay close attention and spend a lot of time in the fields: Carmen puts a thousand miles a year on his John Deere four-wheeler without driving more than four miles from his home.

The Arhuaco had asked if I am indigenous to my homeland. They were disappointed that I wasn't because they imagined all the Indigenous peoples coming together and helping guide our global society to a better relationship with nature. Carmen isn't indigenous to his land either, but he's planted physical and emotional roots that show how building toward such a relationship could be possible for those following in his footsteps.

As connected as he is, Carmen told me, "At the same time, you're so out of control." This got him thinking along spiritual lines. The lack of control, he said, is "why it's so hard for me to give God a gender. There's no way you can define a gender when you are out here. You just can't. It's a being, a force. It's a power. That power, that being, that force is everywhere. It's not just in a building or in a book."

I think the Arhuaco would've appreciated his view of God, and maybe even have helped guide him to an even deeper understanding.

I shared with Carmen how I was becoming increasingly uncomfortable claiming ownership over my twenty acres in Indiana. That the phrase "my land" had started to sound hollow.

"It is a privilege to be able to walk out there and immerse yourself and to become intimate with what's taking place," he said. "When you think about that, and then you think about putting something there to kill it...."

"It's like you aren't honoring that relationship," I added.

"Absolutely," he said.

A-Frame Farm was Carmen's Heart of the World. His work here was both a privilege and a responsibility. And Carmen believed that Kernza could help him and other farmers meet that responsibility.

The tops of the Kernza had started to arc beneath the weight of the seeds. Carmen thinks they're about ready for harvesting. He broke off the top of a plant.

"Can I eat it?" I asked after following suit and picking out the seed with my thumbnail.

"Absolutely," Carmen said.

"It's tinier than wheat."

"Right. But it's almost double the size of what we had in 2011." Just as The Land Institute is working on yield, they're also working on grain size. Carmen is hopeful. "I think it's going to become an integral part of our whole system. Kernza can be grazed as well as harvested." And while the market for it isn't dependable yet, a few years ago Carmen had a per-acre net profit for his field of Kernza of $1,500, which far outpaced his neighbor's conventional corn and beans.

He adds, "The other piece that we're looking at is how the current system helps us in weed management, and building soil health and fertility. It's a work in progress. But at this stage of the game, I see a lot of potential. Our creativity is the only limitation." Kernza isn't the only crop Carmen and Luke grow, but it plays an important role as it outcompetes weeds and its roots work their magic.

## Root and Plow

As I think about what it takes for agriculture to be regenerative, I realize patience and hope are key factors. Regeneration requires long-term thinking—quite the opposite of the nearsighted view of agrochemical conglomerates that are legally obligated to maximize shareholder profit.

Planting something once and letting it grow and return and return again, that's the hope of perennial agriculture. It implies an ongoing force, a regeneration of life that spreads and thrives of its own accord. Yes, that applies to a root system, but it could also refer to a network of farmers and researchers from Wes to Lee and from Carmen to Luke.

Carmen, who won the Rodale Institute's 2022 Organic Pioneer Award for "changing the landscape of regenerative organic agriculture for the better," looked out over his forty acres of organic Kernza, not a weed in sight, nodded, and adjusted his hat. I could tell he was thinking about the weather and when he and Luke would harvest this field.

"We're here just to take care of the land and then to hand it off," Carmen said.

Wes Jackson of The Land Institute put it differently when he wrote: "If your life's work can be accomplished in your lifetime, you're not thinking big enough."

# Rethink

# Part II: Rethink

# Chapter 4
# No Soil at the End of the World

# Kenya

**Regenerative Agriculture Values Life and Soil**

I had never felt more part of nature than when I was worried about being eaten.

"So, like, what do you do with this if a lion attacks?" I asked. "Do I use this stick?"

"The sticks aren't for lions," Leshinka said. "They are for the cows."

I stared at my herding stick and accepted my fate: I was a herder on the Maasai Mara grasslands of southern Kenya, at least for the day, and I protected cows from lions. With a stick. For the cows.

Along with two others, Leshinka and I were guiding a herd seven-hundred strong through the forest and to the river two and a half miles away for their daily drink.

Leshinka directed the other herders to keep the cows bunched, while he and I walked ahead toward the trees. It was a perfect place for lions to hide.

"Before, if we saw a lion," Leshinka said, "we must kill it." Hunting and killing lions was a way for Maasai men to prove their courage and strength, and to protect their livestock. The warrior who held the tail as the lion was speared was deemed

PREVIOUS SPREAD: Regenerative agriculture asks farmers to rethink a system that is reliant on machines that cost more than homes. PAULO WHITAKER / REUTERS

OPPOSITE: Leshinka with his traditional Maasai herding tools, making sure no cattle are missing and looking for signs of lions. KELSEY TIMMERMAN

the most courageous. "But now we don't carry spears," he continued. "We are now friends with animals—buffaloes, leopards, and lions. We don't fear them."

That made one of us. Most of the time, if a lion sees a human, it'll walk away. But sometimes, if a lion has a cub or a kill, it'll get tense, lower its head, lock you in its gaze. There's only one thing to do: stare back and don't move. And then the lion will charge—more than three hundred pounds of muscle, tooth, claw, and evolutionary hunting prowess targeting you. There's still only one thing to do: nothing. Stand there staring at the charging lion and pretend it's a friendly woodland squirrel. There's no point in running. A lion can easily sprint almost twice the speed of the fastest human. So, stand. Since you are such an unafraid badass, the lion will hopefully stop, second-guess itself, and relax slightly. At that moment, you slowly begin to walk backward. Congratulations: you survived.

My heart pounded in my ears at the mere idea there could be a lion in the trees. The lion-colored grass swished against my nylon pants and seeds stuck to the cotton *shuka*—a long rectangular blanket—a Maasai friend had insisted I wear while herding. I had tried to leave it in the SUV that had dropped me off.

I felt uncomfortable wearing it for several reasons. First, the phrase "cultural appropriation" came to mind: I looked like I was playing Maasai. Then there were the colors, which wrapped around me like food packaging advertising the deliciousness within. The plaid *shuka* was pink and purple—flesh and meat—and also black and white—bone and bile. Leshinka wore a dark green winter coat.

He knelt and examined a lion footprint in the mud. It was at least a day old. Encounters with lions aren't out of the ordinary. I watched a mother cow lick a recent lion bite on her calf's head. The herders had chased the lion away and prevented anything worse.

Satisfied, Leshinka hollered to the others.

"Now we push them," he said, and gave a warbling whistle. The other herders led the cows, and we followed behind through the trees, and down a corridor of grass. The whistle communicates to the cows that everything's okay. Maybe it sung to something inside me as well, because my fears evaporated. It was a gorgeous day. The noon sun chased away the cool morning and revealed the abundant megafauna that thrives on the healthy grasslands. Curious giraffes stood at the edge of the woods to our left, slowing their chewing as we passed. Occasionally antelope would bounce above the grass in the distance like fish jumping from water.

"When you look at cows, no more stress," Leshinka said as we walked.

There's something to this. As much as Leshinka calms the cows, they calm him. Domesticated animals do have that effect on us.

The human-animal connection is essential to who we are. Biological anthropologist Pat Shipman argues that the connection had "a fundamental and enormous effect on human well-being." There's evidence of domesticated dogs dating back twenty thousand years, which means we started to rely on each other for protection, food, and companionship thousands of years before we started growing plants. Cows came later, but the connection is undeniable.

As we flowed around the herd, keeping them bunched, Leshinka suddenly looked up and hollered at the other herders, "One is missing!"

I stared at the seven hundred cows in disbelief. "You can tell if one is missing? How?" I asked.

"We have many ways. You are modern so you count one-two-three, but for us...," Leshinka said, trailing off as he scanned the herd.

"You just know?"

"Yes, the white fella is missing. A very short cow with two little horns."

There was more hollering and then Leshinka spotted the tired calf lying on the ground. There would be no truck coming for it. It had to walk on its own to the river and back, or die. If the lions didn't get it, the hyenas, which come out at dusk, would.

"That's it," I thought. "Poor little guy isn't going to make it. It's survival of the fittest, circle of life out here."

But Leshinka threaded his way through the herd, found a certain brown heifer, and walked her back toward her calf. She nosed the calf's butt into the air, nursed him a bit, and then they caught up with the herd. We all made it to the river safely.

"So... cows... how many?" Leshinka asked, turning the tables to ask me a question.

"I have zero cows."

"If you don't have cows, you are not a man," Leshinka said. "You should get some."

I laughed, but he didn't.

We made it to the Mara River without incident and the cows drank. I envied them. The equatorial sun had become unforgiving. I was

## No Soil at the End of the World

dehydrated and had a headache. Normally the herders don't eat or even carry any water, but because I was with them, today we were stopping for lunch, provided by my guesthouse at the Enonkishu Conservancy.

Most tourists visit the conservancy to see the wildlife, but I was here for the cows and to see for myself if the claims that regenerative grazing could fight desertification and undo the harm of industrial agriculture were true. They seemed counter to what I had learned in entry-level environmental science classes in college.

In the Maasai language, Enonkishu means "place of healthy cattle," but that wasn't always the case. This place of elephants, curious giraffes, gazelle, wildebeest, leopards, hippos, countless bird species, lions, and cattle was also a place of corn only a few years ago.

It's as if the giant industrial agriculture machine wants to infest the whole world with corn. And farmers around the world go into debt for the honor of planting and being beholden to the global commodity market. This land, a thousand acres, was owned by the Woods family, who were of European descent and one of the largest grain farmers in Kenya. They were hardcore. Big tractors. Lots of inputs. By 2013 they faced a mountain of debt, interest rates through the roof, and commodity prices through the floor. It was their own farm crisis, like the one Lee DeHaan's family had faced in Minnesota.

If the Woods family did nothing, the bank was going to take the farm. One of their sons, Tarquin, and his wife, Lippa, had a crazy idea: turn the farm into a conservancy by managing it to protect wildlife. Typically, cattle aren't at the heart of conservation

strategies, but Tarquin and Lippa wanted to use regenerative grazing to undo the harm of growing industrial corn. After some initial hesitation, the family agreed—not that they had any better options—and then pitched the idea to other farmers and landowners in the area, big and small, including the Maasai, who had the most experience working with cattle.

They came to an agreement that limits permanent structures and fences, protects wildlife instead of seeing it as a nuisance, and invites participation in the conservancy-managed community herd. As of 2023, the conservancy encompassed 6,000 acres. It generates income by hosting ecotourists and renting land to a farm that produces essential oils. Members share the cost of managing the herd, maintaining a high-walled pen that protects the cows at night, hiring herders, and participating in

Each day the herd is led to the croc-infested Mara River, but neither the cattle nor the herders seem that worried about the potential threat. KELSEY TIMMERMAN

## No Soil at the End of the World

a breeding program. Overall, with the herd at the heart of the conservancy, members make more per acre than they did when they managed their lands and cattle individually.

And it's led to a dramatic environmental turnaround. Even before the Woods family bought their land here in 2000, parts of it had been farmed without any concern for life in or on the soil, other than corn. Like Mark Shepard's little hills in Wisconsin, the soil had been farmed to death. Artificial fertilizers provided all the fertility. Even native grasses wouldn't grow.

Tarquin and Lippa were inspired to work with cattle after learning of the work of Zimbabwean livestock farmer and innovative ecologist Allan Savory, known for his influential 2013 TED Talk, "How to Fight Desertification and Reverse Climate Change." In his talk, Savory, who's redefined land management with his concept of "holistic management," referred to climate change as "the most massive tsunami" and showed how two-thirds of Earth's land was turning from green to brown.

In the 1950s he was a young biologist involved in the creation and protection of national parks and like everyone else, he believed that desertification was caused by overgrazing. His job often involved removing people indigenous to the land, like the Maasai, in order to "protect" it. But the land started to deteriorate. It was Savory's job to figure out why. He pointed at the elephants. Surely their overgrazing was the problem. So in what he calls the "saddest and greatest blunder" of his life, he led a government effort to kill forty thousand of the majestic pachyderms.

Then desertification accelerated. Savory saw similar things happening at US national parks, where setting aside land to let it rewild on its own often led to a decline in its health. Something clicked.

In his TED Talk, viewed by millions, Savory says: "We have failed to understand that the soil and the vegetation developed with very large numbers of grazing animals, and that these grazing animals developed with ferocious pack-hunting predators." To return health to these lands, he said the only option is "to use livestock, bunched and moving, as a proxy for former herds and predators, and mimic nature. There is no other alternative left to mankind."

One natural example of bunched herds rotating across the land is the annual great migration through this area and into the Serengeti involving more than a million wildebeest and hundreds of thousands of zebras, gazelle, and other animals. While the animals cover a vast area, they stay in bunches for protection from predators. And the grasses have time to rest for weeks or months between visits.

Back at Enonkishu, after three years of the conservancy running their herd, the farmed-to-death land around them was covered in grasses. The cattle were thriving, and wildlife was returning. What's more, one area that had had zero springs now had seven. Below ground, the roots of the more abundant vegetation allowed the soil to hold more water, and above ground the shade of the plants kept the ground cool and slowed evaporation. When drought struck in 2015, neighbors lost 30 to 40 percent of their cows, but every single one of the conservancy's cows made it through. Cows, properly managed, brought this land back to life, and the land took care of the cows. While local farmers learned these practices, they also benefited from the predators and other

## No Soil at the End of the World

animals attracted to the healthier savanna—and the tourists they draw. The conservancy has created around eight hundred jobs, from herders to guides to guesthouse staff and rangers.

They were used to catering to people like me. At lunch, they even brought ice-cold beer, which I had to politely turn down multiple times. When we were done, with a storm brewing to the north, we brought the cows back from the river. Set against a dark sky, the yellow grass glowed and waved in the cold wind. The first raindrops were large and infrequent.

"Please," Leshinka said, motioning toward my *shuka*, which he proceeded to tie in a way that it became a coat with a hood. I thanked him. As soon as he was done, the heaviest of the rain began. He pulled up the hood of his coat around his head. It turned out that the *shuka* was quite appropriate for a day of herding in the sunshine and the rain.

These cows were skinnier than those back in the Midwest, but if I focused on a single one and the patch of grass on which it stood, I could be back home. Or more accurately, I could be back to the home of my youth, given that I rarely see a cow—or any farm animal—outside these days. As a kid, my school bus route passed a pasture. To us, cows lying in the grass meant that rain was coming. Science doesn't back this idea, but to us, tiny little humans who hoped we'd get home to play outside before the rain began, it was fact. This connection between the humans, animals, and weather ran deep. The kind of connection that most of us have lost without even realizing it. The kind of connection that may have literally made us human.

The sky was bigger here in Kenya. The forests lower. Standing on the gentle sloping savanna, I swear I could see Tanzania, five miles away. The Great Rift Valley had been cracked open by two diverging tectonic plates. Fissures stretch-marked across the land, which is literally being pulled apart by forces much more ancient than humanity. But there was another rift, a human rift. Humanity and its 'cide-obsessed agriculture was its own force, threatening the life on the land and the lives of humans closest to it. These cows, giraffes, lions, herders, and this conservancy are trying to heal that rift.

•

**Kajiado County, Kenya**

My Maasai friend Dalmas Tiampati smiled as he saw me walk up to the conservancy guesthouse holding the herder's stick, wearing the *shuka* he'd given me.

"Tim! Now you are also Maasai!"

I had him call me Tim, because like many people around the world he had trouble saying Kelsey. The "l" and the "s" are too close for many. That's right, I travel under the alias Tim Timmerman, so more than a few people around the world probably think I sell used cars back home.

Dalmas had been on the phone all day while I had tried my hand at herding with Leshinka. He lives in Nairobi, about 150 miles away, with his three rambunctious boys and his wife, who works for an NGO. He also cares for the children of his sister, who had just died from COVID. He went to college, got a master's in business administration, had a great job at a university. And now he also proudly manages his herd on his ancestral land.

**No Soil at the End of the World**

I accompanied Dalmas from Nairobi to Enkutoto Nalala, his home village a few hours south near the Tanzanian border, on a Friday, an animal market day. The cool breeze whipped clouds of dust that coated the winter beanies worn by many of the buyers and sellers. School fees were due, so it was busy. It was a buyer's market. There were no livestock trailers to be seen; some of the herders had walked for days to bring their animals to market. Many of the purchased animals would be walked away.

Dalmas bought his first cow here for $400 nearly twenty years ago.

"How many do you have now?" I asked. "Or is that an insensitive question?"

"That's a good question, but traditionally people don't want to disclose the number of children they have or the number of cows they have," Dalmas said, explaining that he maintains two separate herds. "In one of the herds I have a hundred and one, and another I have twenty-eight."

Leshinka had informed me a man was rich when he had 150 cows, so I guess Dalmas was on his way.

Two men in winter coats huddled together, negotiating over a brown-and-white heifer. Dalmas told me that some people buy cows just because they're beautiful. He confessed that sometimes he comes here intending to sell a cow, but then can't bear to do it because he likes it too much.

The smaller cows at the market were traditional ones from Tanzania. They held up well in arid climates.

As Leshinka had intimated, a Maasai man's worth is measured by the size of his herd. When he marries, he pays a bride price to his wife's family with cows. A young warrior is rewarded with cows for his bravery. Cows will be taken from an individual as punishment for bad behavior. Peace between tribes is made by the exchange of cows. And they represent the sole source of income for many.

So when, for example in 2009, 80 percent of the cows in Dalmas's village died in a drought, it was much more than a business loss. It was a cultural one.

And then when another drought hit in 2015, his community was devastated again. At the time, Dalmas was working as an administrator at a university a day away, and mothers made the trip to ask him for money for food. He was the only one from the village who had an income outside of cows.

Droughts were becoming much more common. The grasslands and savanna were shrinking, and his wasn't the only Maasai group falling into abject poverty. One of the cruelest things about our changing climate is that the people who've contributed the least to the destruction suffer the most.

Get enough droughts running together and you no longer have droughts; you have a desert. A desert isn't just a lack of water, but a lack of hope. The world's subtropical deserts are growing toward the poles at the rate of thirty miles per decade, supporting less life and turning the world from green to brown, which means fewer plants pumping carbon back into the ground, lifeless soil becoming sand, and more displaced people.

It would've made sense if Dalmas, who'd "made it" by modern standards, turned his back on his herd and traditions and instead doubled down on his career in higher education. But that's not what he did. In fact, the droughts spurred him to quit his job, replenish his herd at the market we visited, and begin searching for ways to regenerate the land.

In 2015, Dalmas came across Allan Savory's TED Talk. Savory's message offered hope.

"That's what I'm going to do," Dalmas thought.

Dalmas emailed Savory, who suggested he visit the Mara Training Centre started by Tarquin and Lippa Woods at the Enonkishu Conservancy. The center acted as a hub of the Savory Institute. Dalmas learned how to properly create and manage a grazing plan, but more importantly he learned about Savory's ideas about holistic management, a system of thinking about the relationships between people, land, and animals.

Eventually, Dalmas even visited Savory at his ranch near Victoria Falls, a place that was strikingly green and alive in comparison to the surrounding land.

"I gave him some Maasai gifts," Dalmas said, and they talked around a fire. "He told me, 'Dalmas, your community depends on you. You are the future of the Maasai.'"

That's a lot to put on a fella, but Dalmas has stepped into the responsibility. "I'm designed to help others," Dalmas told me. He took what he learned and started the Maasai Center for Regenerative Pastoralism, which promotes holistic management to his people in the hopes he could replicate the impact that Savory and Enonkishu have made. I was starting to see that regeneration was contagious. When people saw land come back to life with their own eyes, they wanted to learn about it and spread it to their own communities.

Bomas made from brush protect cattle from predators at night. Even so, their biggest threat is drought. In the drought of 2009, 80 percent of the cattle in Dalmas's village died. LISA HOFFNER / NPL / MINDEN PICTURES

- 

As Dalmas drove us to his house there, he told me about growing up nearby in a traditional Maasai community. Twenty-two family homes encircled a central communal space. There was no individual land ownership. Everyone had a role in village life. Visitors from the outside were to be feared, so if you saw a car, you were told to run.

Ceremonies marked the movement from one role to the next. A boy became a man through the circumcision ceremony, after he had proven he was able to herd large animals and carry a spear for seven consecutive days and nights spent outside in the cold. Then, without any pain meds or anesthesia, he would be circumcised. To demonstrate the Maasai's legendary strength and bravery, it was critical that he didn't flinch, grimace, squint, or react in any way.

Newly minted men were charged with protecting the tribe. One day Dalmas was among a group of warriors fighting a lion.

"For me it was not actually about killing it," Dalmas said. "It was about trying to save two people who were overcome by emotion. I was seeing that they were going to die, that they were just going to be devoured by this lion. Then with one toe, the lion just...." Dalmas made a slash at his leg where he had a visible scar.

I had to pull this story out of Dalmas. It's not something he would share without my asking. If I had lived through something like this, I'd probably walk around in a shirt that read: "Ask me about my lion scar."

But Dalmas just added, "Finally, we speared the lion."

## No Soil at the End of the World

Dalmas's work is supported by Soil4Climate, a US nonprofit that promotes and advocates for "soil carbon activists." Seth Itzkan, one of its founders, sits on the Savory Institute's board, and has visited Dalmas in Kenya, seeing the importance of helping the Maasai, who he said are on "the forefront of the effort to restore soil carbon to reverse global warming." The idea is that if Dalmas could succeed here, in an area even more arid than Enonkishu, success was possible anywhere.

Dalmas learned to herd as a child, starting with sheep and goats. Each night the adult herders would bring the cows back to the village, and the community elders would discuss where to graze next. They would predict droughts by observing the stars, winds, cloud formations, and animal behavior. Essentially their culture of pastoralism was centered around rotational grazing for hundreds of years before Savory's model of holistic management and terms such as "regenerative agriculture" were circulated. In fact, Savory credits the Maasai and other Indigenous people as important influences. For Dalmas, today's holistic management recalls an aspect of Maasai culture that has been lost.

One day in the late '80s, when Dalmas was a boy taking care of calves and lambs, the Kenyan government official came to his village and informed its residents that their community land had been subdivided. There was nothing to be done about it. Dalmas's grandfather was allotted fifteen hundred acres, which were eventually divvied up among his many children from several wives. Dalmas now owns 102 acres.

That was the end of communal living and its rich traditions. The tribe's herds were no longer managed together, bunched, and rotated. Cows grazed on the same land day after day, giving the grasses no time to recover. The land began to die, and in less than one generation, gazelles, giraffes, elephants, and rhinos vanished from the area. There were still Maasai warriors, but now there were no lions.

The degraded land would support fewer cows. Needing an income, families cut down their trees to sell as charcoal in the city. When there was no other way to support themselves, the Maasai sold their land, moved to the city, and did their best to buy food and charcoal from others.

"Capitalism is not the best way for the Maasai," Dalmas said. "We were meant to live together."

•

Dalmas took me to the place where he lived as a boy. Two large hills capped with rocky outcroppings overlooked the site.

Fine dust coated the inside of my ears; plumes of dust accompanied each of our movements. It hadn't rained in eight months.

A door made from wooden planks hung in the frame of two mangled, leafless trees. It was the only entrance in a fencerow created from brush with protruding thorns as long as fingers. Dalmas's communal childhood home, made of mud and thatch, once stood behind it. His family and friends—his everything—had lived there, and now there was nothing. No homes, no people, no animals. Clumps of young acacia trees, not yet harvested for charcoal, clung to life.

And yet Dalmas could still picture what this place once was and hear the echoes of the past. He pointed to the area where the

Maasai children care for animals too young to be led to grass and water during the day. Cows play a critical role in Maasai culture. They change hands as a bride price and to settle conflicts. KELSEY TIMMERMAN

cows were often homed at night. "When it rains, this is the first place to get green."

There was more grass and more green in that area. After all these years of neglect, the positive impact of the bunched cows from decades ago was still written on the land amid the evidence of smaller, individual herds overgrazing indiscriminately.

A few years ago Dalmas attended a meeting convened by the United Nations. After Allan Savory spoke, an attendee stood and disagreed with everything he had said. She said that pastoralism must be stopped. It was the old argument that cattle and pastoralism, regardless of management, led to desertification,

and that the people who lived on the rangelands and pastures that cover half of Earth's landmass were the destroyers of their own homelands.

"She wants the Maasai to disappear," Dalmas thought. "When you destroy pastoralism, you destroy our culture."

Since 2009, the Tanzanian government has displaced one hundred fifty thousand Maasai. In 2022, Maasai just across the border from Dalmas's village protested yet another eviction and were met with rubber bullets and tear gas that injured thirty people. In the name of conservation, borders are set, the Maasai are forced from their lands, and tourism and hunting permits are issued to wealthy visitors. This antiquated path to conservation from the colonial past continues to this day.

The door that once opened into Dalmas's communal home now leads nowhere. The only other remaining signs of their traditional way of life here are the green patches of grass where the community bomas once stood. KELSEY TIMMERMAN

## No Soil at the End of the World

"The grabbing of local Indigenous peoples' lands is underpinned by a war on sustainable and self-sufficient ways of life that has been waged for generations," writes Stephen Corry, the former CEO of Survival International, which fights for the rights of Indigenous people around the world. "In the specific case of Tanzania, the land theft is to facilitate trophy hunting by United Arab Emirates nobility as well as for tourism; in the end it always comes down to money and control, not conservation."

And yet, in 2018, Dalmas attempted to replicate the community-herd model practiced at Enonkishu in his village. He sold his neighbors on the idea. Told them to imagine three stones of a cooking fire. The stones represented land, people, and livestock. Without all these stones, they couldn't regenerate the land. They bought into the idea, but before long they pulled out, disappointed to find that Dalmas wouldn't pay for other projects, such as digging wells, on their own land. Then it was just Dalmas and his cows.

Still, he remains optimistic, certain that once they saw how the land, the cattle, and his family all thrived, they'd rejoin his effort. The land that Dalmas manages is greener and more alive than his neighbors', though the drought persists. At times he's had no choice but to make the costly decision to purchase and bring in hay to feed his herd, an option most of his neighbors couldn't afford.

"I think we can do what I do to be sustainable and profitable and can be able to live a good life. Of course, my wife sometimes says, 'I support what you are doing, but do you think we have a future with this?'" he said, pausing before he continued. "I think she's right in having those concerns because it's about us and our children."

Dalmas paused before he told me that some old friends don't speak to him any longer. One, a lawyer, had told him, "Dalmas, you are out of your mind. How do you live in the city and want to go and kick cows?"

"So for him and maybe others," Dalmas said, "they think I'm not okay. But I want to disprove these people."

As we talked, a group of women approached on the road. They were dressed in full Maasai regalia—intricate beaded bracelets and necklaces and earrings, brightly colored *shukas*. It was as if the landscape was dusty and monochrome because these five women had absorbed every crisp, clean color.

Some of them were old enough to remember when the community all lived behind the door. With Dalmas, they remembered the time a cow had fallen from one of the cliffs.

They continued down the road and Dalmas got quiet. He stared up at the hill. There used to be a flag on top. He thought it was a British flag.

"Did you climb up there?" I asked.

"When I became nine years old, we snuck up there," he said. "Those days we were thinking that when we reached the top, we would be touching the sky and touching God."

"What else do you remember?" I asked.

"There were so many kids to play with," Dalmas said. "So much joy in the village. That was a very good life. We had plenty, actually. We had milk in the village. People just singing. We didn't have this struggle of taking animals to find grass."

# No Soil at the End of the World

"Can I take your picture here?" I asked.

"Let me get my *shuka*," he said.

Dalmas opened the car door and a cloud of dust drifted down the road in the direction the women walked. He put the *shuka* over his black Henley shirt and blue jeans. He put his hand gently on the acacia tree, as if upon a child's head. The door to nowhere was behind him. He didn't smile. I couldn't decide if he was sad or proud.

•

## Kisumu, Kenya

Since the Green Revolution in the 1960s increased yields through developments in agrochemicals and new seed varieties, the United States has exported its input-intensive, commoditized agriculture to the rest of the world, changing landscapes and cultures. Dalmas and I traveled to western Kenya to see the devastating effects of what some are calling the Green Revolution 2.0, which is promoted throughout Africa by various businesses, governments, and NGOs.

This trip brought up a question I never thought I'd ask: What happens when farmers no longer see value in their land and just give up? For the most part, the people I met on my journey have hung onto the hope that we can regenerate our land and each other—but nothing could have prepared me for what I saw in western Kenya, where a young activist named Celestine Otieno showed me the answer.

•

Celestine sat next to Dalmas and me on a thin-cushioned couch in the home of a woman named Susanna. Susanna was happy to see Celestine. Everyone was always happy to see Celestine. I had spent the last few days visiting farmers with her, and she was greeted with hugs, smiles, and respect. Celestine has a degree in agriculture and pursued further education in permaculture. She helps guide farmers like Susanna and advise groups on community gardens, home gardens, and agroforestry projects. Celestine has a laid-back way about her, and her enthusiasm and energy never flagged, especially when compared to Susanna, an exhausted mom of eleven.

Chicks made their way in and out of the front door. Five bags of corn were stacked in the corner of the home.

Susanna had been sick for some time, but didn't say what her illness was. Just that she had to have help with the harvest. The largest portion of her land was dedicated to growing corn.

Celestine walked me through the economics of the current growing season. An international NGO had given Susanna corn and fertilizer in the form of a loan at the beginning of the season. She grew one acre of corn, which yielded the five bags in the corner. Her plan was to keep one bag for her family to eat and sell the other four, though the prices were low because everyone had just harvested. But she had to pay back the loan, valued at $88. She'd probably get around $18 per bag, $72 total, for the four bags, so she had to find some other way to make up the difference. After an entire season of growing corn, she'd be $16 poorer. And thirty days from now, when her family had eaten their bag's worth, she and the other farmers in her position would be buying corn at the market, where the price would be double what she was paid.

"For every season after one month, they have no food," Celestine said.

Dalmas, a man between two worlds, was educated in a university, but now attempts to reconnect his rural neighbors by putting their cattle in a community herd. Many of his city friends think he's wasting his time and talent. KELSEY TIMMERMAN

This was Susanna's fourth season growing corn with the seeds and fertilizers provided by the NGO, and she'd yet to make any money. My farm friends in Indiana and Ohio could relate to this situation. The difference was that my friends back home weren't subsistence farmers already on the edge of malnutrition and extreme poverty. Farmers in the United States are supported by crop insurance, subsidies, and a social safety net. Farmers like Susanna aren't.

Susanna used to grow beans and ground nuts, but since she started growing corn she couldn't afford to buy those seeds any longer. She owned four acres and didn't have the resources to farm all of it.

I didn't understand why she and the many other subsistence farmers Celestine introduced me to even bothered with corn. Not one of them was making money on it.

When I asked, the answers weren't unlike what I hear back home: "We grow corn because that's what people do around here." Being able to access the seeds and inputs on credit was also critical to their decision.

While it clearly wasn't working, the predominant hope of the international community of funders was that this type of agriculture would lift people out of poverty. Imported seeds controlled by corporations replaced local seeds. Monocrops of an imported plant replaced diverse fields of native foods.

This was happening across Africa. A case study in Malawi, for example, found that after heavy investment in industrial agriculture, the poverty rate remained the same and food insecurity increased slightly.

## No Soil at the End of the World

One of the biggest supporters of industrial agriculture in Africa is the Bill and Melinda Gates Foundation. The foundation is the primary funder of Alliance for a Green Revolution in Africa (AGRA), which from 2007 to 2018, for example, provided $22.5 million of funding to support the sales of commercial seeds, fertilizers, and pesticides. Not everyone welcomed the support of this type of agriculture. In fact, around the globe, subsistence and agroecological farmers and organizations such as GRAIN, an NGO that supports small farmers, have spoken out against their efforts.

In Africa, the South African Faith Communities' Environment Institute (SAFCEI), which Celestine was a member of, wrote an open letter to the Gates Foundation pleading for support of regenerative practices. The letter expressed concern about the environmental consequences of chemical-intensive agriculture, the fact that farmers were growing less diverse crops and suffering nutritionally, the support of big agrochemical companies via the distribution of GMO, or bioengineered, seeds, and the possible expansion of Cargill's soybean market "with a damaging record in South America."

SAFCEI's executive director was interviewed about the letter and added, "African farmers need communal solutions that increase climate resilience, rather than the top-down, profit-driven, industrial-scale farming systems that the Gates Foundation is supporting."

The reporter also interviewed Celestine. "Farmers in my region are becoming wary of programs that promote monoculture and chemical-intensive farming," Celestine said. "We are losing control over indigenous seeds and farming systems and are being held hostage on our own farms. The Gates Foundation is pushing to expand industrial agriculture," she said. "My question: Is agricultural industrialization leading to food security or food slavery?"

The Gates Foundation never responded.

"Will you want your kids to be farmers?" I asked Susanna.

She strained to laugh and in the wake of the laughter, her voice grew weaker. "It's a hard life I wouldn't want for my kids."

As we went outside, she walked beneath a poster of a plant emerging from dark, rich soil. It was hung upside down, the plant growing from the ceiling, and underneath it was one upside-down word: Hope.

Hope is upside down when it must be bought, when fertilizer must be mined from far away, and when seeds are owned and shipped around the world, entrapping people like Susanna.

•

Our next stop was a very different garden, a permaculture project that Celestine had been working on with a group of women she called the Apuoyo wives. (*Apuoyo* means "thank you" in the local language.) The group of twenty-four women were all widows, except for Susanna.

"This was my main worksite," Celestine said, stepping from the dirt road toward a large garden of raised beds patchworked with ditches and walking paths. She lifted her arms to the sky and her fingers became rain landing on the ground demonstrating the flow across the landscape she had designed and helped dig. She stepped through the gate in a pea-wrapped fence put

## No Soil at the End of the World

in place to keep the neighborhood goats out and then stood beneath the shade of banana plants.

"First, we had the fence. And then we had the climbers on the side," she said pointing at the peas. "Then we had the sunflowers to attract the birds that would eat the aphids. And then we had the African marigolds," she said, indicating the faded flowers. The marigolds attracted pollinators. The diversity of plants in and surrounding the garden attracted a diversity of insects and birds. Insects provide up to $500 billion of value to global crop production, and yet their populations, threatened by monocultures, agrochemicals, and deforestation, are declining at the rate of 9 percent per decade.

Our best protection against a changing climate is diversity of seeds, of crops, of cultures, but we've chosen uniformity. There are nearly three hundred thousand edible plants on Earth, but only three—rice, corn, and wheat—account for 50 percent of calories consumed by humans.

Instead, Celestine is working with nature's diversity and finding a way to care for the earth while caring for people. Call it permaculture, holistic management, or regenerative agriculture—they all embrace diversity, accommodate the needs of human, animal, and plant communities, and reject industrial agriculture.

Celestine explained the relationships between the plants and animals and the soil. How they each care for the other. Most of the time when someone shows you their garden, the conversation is about when something is ready for harvest and recipes, of what the garden has to offer humans, but it takes a while for Celestine to get there.

"And then we have the amaranth—" she said and then paused. Amaranth is a tall perennial plant topped with red plumes. Its leaves, run with red veins, look like chard. To industrial agriculture, it looks like a weed to be eradicated. But consuming it can be healthful for a population at great risk. "I'm sorry to say some of these women have a low immune system because of HIV and they can benefit from eating amaranth porridge," she added.

The widows, who all lived nearby, tried to gather at the garden to weed and harvest once a week. They took some of the vegetables home, and the rest were sold at a local market and to the local hospital and school. A proud woman wearing a purple T-shirt, a matching purple checkered wrap around her waist, and a green bonnet joined us. Ann, a member of the group, had donated the area for the community garden.

"When we first started we made only two hundred to three hundred shillings [$2–3] per week," she told me.

"Last year we had carrots, cucumbers, strawberries, kale, [capsicum, onions], and beet root," added Celestine. "We could sell five thousand shillings [$45] per week." Celestine hoped this kind of growth would encourage the women to learn, or in some cases remember, permaculture principles in the community garden and then apply them on their own land.

We turned from the group project to visit Ann's private garden. A variety of vegetables in raised beds cut with ditches and walking paths was next to a plot of scrawny corn, for which she had received a $65 loan from an NGO in the form of seed and fertilizer. Celestine rubbed her hand across the top of a knee-high cassava plant, growing between the rows of corn. "She did this against the [NGO's] will."

## No Soil at the End of the World

The corn wasn't producing enough to pay back her loan, so like Susanna, Ann was falling behind. She hoped that the cassava could catch her up next year, allowing her to pay off her debts and leave her with some food to eat. This was her second round with the NGO. This year she had sold a goat and a few chickens to make up the difference.

"I want to plant more vegetables because within two months you can take it to the market to get income," she said. "I just want food to feed my kids."

"Before you started growing cassava and other vegetables, a bad corn harvest meant you wouldn't have any food to eat?"

"Yes," Ann said. "We totally depend on farming."

"When the deal goes wrong," Celestine said, "the vegetables are the back-up plan."

"What can we do next, so we can move forward?" Ann asked me as we stood on the path dividing the two gardens, the two different ways of agriculture. Surely someone who looked like me, who'd come from so far away to meet her, had some answers or advice.

I laughed awkwardly. The problem has been that people like me who come from where I come from think we know what's best for people like her. I don't have any agronomical advice, but I'm pretty sure I had the correct response. "Celestine is the expert," I said.

•

"Tim, be sure to drink lots of water. It is hot and dry here," Dalmas said.

"So where are we going? To see sand?" I asked.

"Yes," Celestine said, "Sand harvesting. Near my home." She'd grown up in a nearby village, surrounded by lush farms and forests that stood out against the rolling hills of dusty brown soil.

I was clueless. I didn't know what sand harvesting was or why it was relevant. I hadn't ever thought much about where sand came from.

We left the car behind one of the few buildings that made up the commercial center of her village and Celestine led us down a road so rutted and gullied that only a truck could manage it. On our right, the ground sloped up and then suddenly, there was nothing. The hill was gone.

The amazing Celestine on the right, with one of the women she taught to garden using the principles of permaculture. Celestine's work provides an alternative to industrial seeds and inputs, which often saddle farmers with unmanageable debt. KELSEY TIMMERMAN

Sand harvesting. It looked like a bomb had been dropped, and all that was left was a crater the size of a house.

"This is bad," Dalmas said.

Next to water, sand is the most consumed natural resource. Each day forty pounds of sand are extracted for every person on Earth. Sand is the largest traded solid resource by volume, globally. In most places, including Kenya, the industry is unregulated and run by organizations referred to as "sand mafias." Sanding feeds the endless need of the construction industry, where it's used to make everything from glass to cement to tarmac.

Sometimes sand comes from coastlines or riverbeds. Other times it's taken out from under what used to be a farm.

It felt dangerous being there.

Celestine agreed, leading us farther down the road. "Be careful. Some sand miners use drugs and become quite aggressive."

And it felt sad, too. We passed a man wearing a tattered suit coat and pants, a short-handled hoe hanging from his shoulder.

Celestine recognized him and told us, "They discovered that his sand was so fine. Trucks were trooping in his home day and night.... Fifty trucks per day, so he used to get a lot of money. He became wealthy in a day."

"And then it all disappeared?" Dalmas asked.

"Yeah," Celestine said. The sand ran out. "And now there's no more tillable land. So his son... where will he build a home?"

A younger man approached, Frederick, a former classmate of Celestine's. He led us along a path that skirted the edge of a house with a small farmyard. Forty feet behind it, the land ended in another sand crater wasteland, this one enormous, the size of New York City's Central Park. A burst of gritty, hot air whipped our faces, sticking to my sunscreen. We had to turn and squint to keep the sand out of our eyes.

At the bottom of what used to be a hillside, people with hoes scraped at walls while others loaded the trucks. I thought about a senior UN official, who warned in 2014 that we only had enough soil for sixty years of growing seasons. This must be what year sixty-one would look like. The place felt like an active crime scene. A crime against humanity and nature.

"This was an unproductive corn farm," Celestine said. "These farmers have lost hope with farming. So the only thing that would give them quick cash is harvesting sand."

We walked at the pace one walks in a graveyard. Celestine and Frederick waited for Dalmas and me beneath the shade of a spared tree.

"Do you see those strange pieces of land?" Celestine said. Towers, sometimes twenty feet tall, were all around us.

They looked like those sandstone towers at Arches National Park. Except the towers at Arches took sixty-five million years of wind and water to form. These ones were formed since 2018 by a labor force of poor farmers who could no longer make a living.

Some of the towers were topped with toupees of topsoil, which still fed life to grass and trees.

"Those are graveyards. The miners cannot touch them," said Celestine.

## No Soil at the End of the World

"This is a graveyard," I echoed.

The Luo people of Kenya have very particular burial traditions. For instance, if an unmarried man dies, he is buried on the right side of his mother's house. If a woman with children dies, she is buried on the left side of her own house. Now there are no houses, just towers, hundreds of towers, hundreds of graves where generations of people had been buried.

"I think they should've left alone the whole area," Dalmas said.

"How many people have died so far working in the sand?" Celestine asked.

"Here?" Frederick asked. "Seven."

"When was the last time someone died?" Celestine asked.

"In January," he said. It was now August. "Morris." Celestine looked shocked.

"You knew him?" I asked Celestine. She nodded.

Sometimes the workers were asked to dig in trenches or even in caves that collapse. That's what had happened to Morris.

"What is causing this?" Dalmas asked.

"Poverty," Frederick said. "Sometimes we have no money for food or to pay the school fees of our kids."

After a while, we turned our backs to the miners and walked up the hill, stopped at the farmhouse on the edge, and took it all in.

"This is what happens after planting and using fertilizer," Celestine said. "This was basically all hybrid [NGO] corn."

In shock, feeling hopeless, I asked what could be done. And Celestine, who never seems to give up, had an answer.

"First, community awareness and valuation of our properties," she said. "And then we really need to understand the dynamics of nature through the Indigenous knowledge, and in that way we can be able to restore our land."

Dalmas nodded his approval.

"Do you think if hybrids hadn't been introduced, and monocultures and modern agriculture hadn't been introduced, that people would be able to make a living as farmers?" I asked.

"Yes. For sure. That's what I guarantee you," she said. "Was I telling you how different even the harvests were when I was growing up? The crops were so different from the crops we get now. My grandmother had a whole different practice. She would just scratch the land. There was no intensive cropping. We traded. I come with my maize, you come with your sorghum or fish. I give you maize, you give me fish. They give us, we give you.... So I really feel if we still had that tradition, then we would be so stable."

An empty sand truck came toward us with diggers on the back. They were refugees of industrial agriculture now employed in the further destruction of their community, and our world. Paid to dig canyons on land passed down from the ancestors they now dug around.

Do they still visit the graves of their ancestors? Instead of mourning toward the soil, do they stand at the base of the grave towers and look to the sky? Sanding seemed to kill both the past and the future.

## No Soil at the End of the World

We stood in the way of the truck: Dalmas, a Maasai warrior who knew his culture had a solution; Celestine, an anti–Green Revolution activist who believed in the regenerative power of permaculture; and me. I wanted to hold up my hands and yell, "Stop!"

But even if I prevented this one truck from reaching its destination, there were many more. We stepped to the side of the road and the diggers stared at us as the truck passed.

Celestine waved to them, and said, "If something specific was introduced to them, they would redirect that energy to a more sufficient project."

She outlined how she would mobilize and educate her community to restore their land, how she hoped to further pursue her education and ultimately develop a training facility in her community to offer an alternative to industrial agriculture and sanding. Dalmas asked Celestine a few questions about permaculture. He was interested in helping women in his own community start gardens and food forests.

When I looked at the sand crater, I saw the end of soil, the end of our world as we know it. But as I listened to Dalmas and Celestine, I realized that the sand crater motivated them, and was reminded of how important their work was.

Soil is life. Sand is death. A world where sand is worth more than soil is a dying world. There are places on Earth where the world is literally ending. But there are still people in those places who believe their neighbors, and their communities, can heal it.

# Part II: Rethink

# Chapter 5
# Eating the Rainforest

# Amazon River Basin, Brazil

**Regenerative Agriculture Protects Our World**

Smooth clouds reflected the texture of the river. The sound of the ferry's engine thumped over the banks of brush and trees too short to return an echo.

I was alone on the top deck of the *Amazonas*, relishing the view without the distraction of the ear-piercing Brazilian pop music I knew would start up again before noon. Emptied one-dollar beers sat on wooden tables and plastic deck chairs. Last night's drinkers swayed in hammocks on the second and third decks. The ferry was industrially functional, a giant floating rectangle. The shared showers and toilets were rusty steel coffins.

At a dock in front of a house on stilts, a man pull-started an outboard engine on the back of a dugout canoe. The ferry slowed, holding its own against the current. The man tied off to our port side and helped an elderly woman aboard. The ferry was a bus and people were constantly entering and exiting.

I monitored our progress by the GPS on my phone as we made our way into the Brazilian state of Pará and deeper into the Amazon Basin. Plants with pink flowers and leaves like cobra heads wiggled in the current as if the shore were alive. Occasionally a river dolphin would surface, but overall, I was surprised to see so little actual wildlife. There weren't many birds or fish. By far the most common animals were the cows standing

The *Amazonas* serves as bus and delivery service up and down the Amazon River, passing barges of soybeans grown where rainforest once stood. KELSEY TIMMERMAN

in large pastures of cleared forest, and even at times standing in island corrals completely surrounded by wetlands and water—odd floating CAFOs. I saw more barges than dolphins, transporting load after load of soybeans.

The pink sunrise faded to gray as the clouds inhaled the sun.

As I stared at the passing wetlands, black greenhouse diesel exhaust belching from the ferry's smokestacks, it was hard to get climate change out of my mind. I thought of machines around the world with valves, pumps, dials, and computer chips dedicated to gathering carbon and pumping it back into the ground. I remembered the climate solution chart a champion of Kernza had shown me back home, describing the potential of direct air capture (DAC) machines. They're pretty ingenious, but one scientist described their current capacity to sequester carbon as "trying to bail out the *Titanic* using an eyedropper."

There is an older, cheaper, more efficient, and more complex technology: Forest. A patch the size of a large Midwestern family grain farm captures as much carbon per year as all of the DAC machines in the world. And while we are investing in them and other technologies, we continue to cut down the rainforest.

In the twelve months before my visit, more than 2.5 million acres of Amazon rainforest were destroyed. That's more than three football fields of forest gone every single minute. And one of the main causes of that destruction? Soybean farms. I'd traveled thousands of miles to find that once again one of the two crops that surrounds my home in Indiana lays waste to its environment.

When I was a kid, I imagined donning a RoboCop suit and traveling to the rainforest to take on the "bad guys" destroying it. Now,

## Eating the Rainforest

in my early forties, chugging up the Amazon, I was on my way to meet the people who might have the best way to do that—using regenerative agroforestry to restore and protect the forests and producing food and income, while sequestering way more carbon than some machine.

•

César De Mendes held out a brick of chocolate covered in thin red foil. "Gold," he said. It was 4:15 in the morning, and I couldn't match his enthusiasm for chocolate or consciousness.

De Mendes is larger than life. He has been called "the Indiana Jones of chocolate" by some, and "the Don Quixote of chocolate" by others. But most people simply call him by his last name. His company, De Mendes Chocolate da Amazonia, supports the production of native cacao by traditional and Indigenous farmers in the Amazon, ultimately bringing chocolate bars to the market. The Dutch NGO reNature, which supports and promotes his and other regenerative agroforestry projects, estimates that De Mendes's work preserves more than 550,000 acres of rainforest and impacts the lives of over 3,000 farmers.

He and his wife live in a small house on a seven-and-a-half-acre homestead in Santa Bárbara do Pará that they've transformed into a food forest. When I'd arrived the previous day, he said, "This is our supermarket and our pharmacy." Think of a fruit, think of a vegetable, and De Mendes could more than likely walk you right to it. He grows nearly a hundred species for research and their own use.

The blue flame of the gas stove licked the silver pot, and the kitchen smelled like a chocolate factory. De Mendes poured honey from his beehives into the pot and then filled porcelain cups with the concoction. Bubbles surfaced and burst with small puffs of aromatic steam. I took a sip. It was somewhere between a liquid and a solid, between heaven and earth, warming my extremities and my soul the way a shot of whiskey would. I'm not sure what my face did, but De Mendes smiled and laughed and took a sip from his own cup.

"I grew up drinking chocolate and still have it every day," De Mendes said. His mom would collect cacao pods from the trees—which are native to the Amazonian forest—ferment the beans for days, and smash them into powder with a mortar and pestle before mixing it with water and honey or other spices and flavorings. Before there were bars of chocolate that could melt, cacao was the key ingredient in drinking chocolate. The Olmec of present-day Mexico, who refer to it as the fruit of the gods, are often credited as the first culture to value cacao, but it originated and was likely domesticated along the equator in the Amazon 5,300 years ago.

Before his current gig, De Mendes worked as a chemical engineer, professor, and entrepreneur. He has specialized in environmental science, food quality, food production, and (of course) chocolate. In his twenties he became interested in philosophy and religion, and told me that exploring a religion was like trying on a pair of shoes. He had tried on and outgrown twenty-eight of them by his count—Buddhism, Hare Krishna, various African religions, Catholicism—before landing on Orthodox Judaism, which clicked, especially after his dad revealed that the family, originally from Morocco and Portugal, had Jewish roots. He wore a black yarmulke and a Star of David on a chain that fit so tight around his neck it made me swallow. It was a little unclear what

his relationship was with religion now. But it was clear he had a mission. He still searched.

The chocolate we drank came from cacao grown by a farmer named Xiba, who lives a half-day journey away, along the Tocantins River, a tributary of the Amazon. We were up so early because we were going to pay him a visit, and in the Amazon Basin you give yourself as much time as possible to get where you need to go. Ismael Nobre, a biologist who works with the Amazon 4.0 initiative—which helps people who live in the rainforest generate income from farming the forest regeneratively—was coming with us to meet Xiba and check out his operation.

César De Mendes examines the food forest surrounding his home. His efforts to promote regenerative agroforestry have been credited with preserving more than 550,000 acres of rainforest.
KELSEY TIMMERMAN

If all went well, it could lead to grants for things like processing equipment, refrigerators, and Wi-Fi.

After a ride in a wooden boat, we arrived at Xiba's place by late morning. "In Pará," De Mendes said, "the river is your street."

Xiba, a man in his forties dressed in long athletic shorts and a polo shirt emblazoned with a crocodile, greeted us, and we followed him from the river through the mangroves to his house. Plants grew in every sort and size of plastic container and jug, lining the walkway, attached to the wall, hanging on a railing, and perched on the kitchen table. Xiba was surrounded by the forest and yet he still brought the forest into his home.

Without further ado, he plunked down two cacao pods on a workbench before De Mendes, who put them to his wide nose and inhaled deeply. He didn't blink and his eyes didn't move, as if directing all computing power to his nose and whatever software processed the incoming data. Xiba waited.

They had first met in 2014 when De Mendes was a consultant on a university project to help farmers develop chocolate-quality cacao and collectivize their efforts. Students and farmers had gathered at De Mendes's house.

Someone gave De Mendes some cacao to taste and as soon as it hit his tongue, he spit it out. "I would never buy this," he'd said, not realizing that the man who had produced it, Xiba, was standing right in front of him. "I wouldn't give this to animals."

Undaunted, Xiba asked De Mendes for help. De Mendes spent nearly three weeks with Xiba's family, working with them on the cacao and building a warm relationship. He guided Xiba and his two brothers through the critical fermentation process, where the cacao beans start their transformation into chocolate. Their biggest challenge was maintaining a steady temperature in a place where the thermometer registered nearly one hundred degrees during the day but could drop to sixty-five at night. The bacteria necessary for fermentation needed the heat to work their magic. De Mendes advised the family to put the fermenting beans inside a black fiberglass bin and put the lid on to keep them warm at night and take the lid off during the day.

De Mendes left and when he returned two months later, Xiba shared a story of how a horrible smell from the bin had woken his entire family one night. In the morning, he took off the lid and put the beans in the sun, hoping to save them. De Mendes, intrigued, took the beans back home to try them—and found that they produced the best chocolate he'd ever tasted. The accident—which he learned to replicate—had produced a flowery taste.

"Normally there are seventy-eight microorganisms involved in the fermentation of cacao," De Mendes said, "but in the Amazon there are one hundred and fifty." This complexity can lead to extraordinary chocolate.

But first the wild cacao itself needs to be of top quality, and that depends on five factors: quantity of water, soil characteristics, amount of light, the type of surrounding plants, and their diversity. Some of this can be controlled, but every area, tree, and farmer is unique and works with a different mix of microorganisms.

De Mendes talks about these details the way winemakers talk about terroir, how the specific character of a wine comes from the place it's produced.

Xiba's front porch is lined with tools to farm the rainforest and live on the river. His style of agroforestry is profitable and highly effective at sequestering carbon. **KELSEY TIMMERMAN**

**Eating the Rainforest**

Character and taste can be hard to standardize across a range of producers, so De Mendes has stopped trying. Now, instead of trying to mold plants, people, and processes into a uniform bar of chocolate, he spends his time seeking out traditional and Indigenous people who've had relationships with native cacao plants for generations, and allows their uniqueness to shape the chocolate.

That search has taken him deep into the rainforest, sometimes traveling with anthropologists who had longstanding relationships with the Indigenous people who live there. Tribes were so leery of the outside world that they shot arrows at passing planes, but they hosted De Mendes. On one fifty-day journey, the river swelled with rain and he and the five men he was with had to camp in the forest. In the middle of the night, he woke up and saw a shadow sniffing him. It was a jaguar.

"I smelled so bad the jaguar ran away," he joked.

As we talked, his phone pinged text message after text message. He was receiving updates from the Indigenous Yanomami. He produces chocolate from their cacao, along with the cacao of another Indigenous group, the Ye'kwana. Cacao provided the groups with an income, allowing them to push against and operate in a world that was invading their lands.

As in so many other parts of the world, the Indigenous people of Brazil are under attack. Within hours of taking office in 2019, former president Jair Bolsonaro transferred the oversight of Indigenous territories to the agriculture ministry. He saw the Amazon as a resource to be exploited, and painted those who would protect it as enemies of the Indigenous people who live there. He wrote that environmentalists and foreigners "want the

Indigenous people to carry on like prehistoric men with no access to technology, science, information, and the wonders of modernity. Indigenous people want to work, they want to produce, and they can't. They live isolated in their areas like cavemen." When he shared his deplorable and racist views on Facebook Live, he stated that the "Indigenous are increasingly becoming human beings just like us."

It's hard to fight capitalism and the point of view that the rainforest and Indigenous land are a blank canvas that don't contribute to the country's GDP. But people like Ismael and De Mendes hope that regenerative agroforestry can help. It can keep people on their land, and by keeping the land "productive," keep the extractors at bay.

"Cacao is the fruit of our forest," a Yanomami leader had told De Mendes, "and a cry of resistance against the invasions that Indigenous lands suffer daily in Brazil."

"They have a very different way of seeing life, which ends up being translated into the taste of the chocolate itself," De Mendes said, describing their culture in a way that reminded me of the Maasai and Arhuaco. "The Yanomami and Ye'kwana are very respectful, sensitive, humble, and humanitarian," he continued. "They do everything as a group. We all have to learn from this posture, attitude, and example of cooperation, solidarity, and looking at the other. They work with a sense of empathy."

For their efforts, the farmers he works with earn six times the going rate for their cacao. In the global market, a cacao bean is a cacao bean, valued more by weight than quality, as it joins a river of other beans grown by farmers whose identities and lives matter less than the flow. But De Mendes is de-commodifying

cacao, helping the world enjoy its unique characteristics and the characters behind it.

I observed that cacao originated in the Americas, but now West Africa is the largest producer.

"When the Spanish colonized the Americas, they brought cacao to Europe as a drink," De Mendes said. "When it came back, it was a bar. They removed the humanity. They don't respect the seeds. It's totally different. Worse yet, they took it to Africa and brought it back with the flavor of blood."

The blood of slaves, exploited farmers and workers, and nations is precisely the flavor I try to avoid. I gave up eating mass-produced chocolate years ago after learning and writing about the cocoa industry of Cote d'Ivoire and Ghana.

Turning the insides of a cacao pod into a delicious and consistent chocolate bar requires a complex fermentation process and careful processing. De Mendes credits the quality and uniqueness of each bar to the character and personality of the farmers who grow it.
JAMES KENDI / GETTY IMAGES

To take us from his house to his farm, Xiba guided his boat up a narrow stream surrounded by jungle. The water changed from a milk chocolate to a transparent tea steeped in organic matter. When he cut the engine, the silence was immediate. I closed my eyes, and what I heard surprised me. Even here, the frogs and the insects seemed to sing the same song the frogs and insects in Indiana sing on the first warm nights of spring.

With the bow of the boat stuck into the bank, Xiba helped us off. Two-by-fours laid end-to-end created a walkway across the squishy forest floor, which led to an elevated narrow bridgeway constructed with hammer and nail. The natural food forest surrounding Xiba's intentional food forest was overflowing with a lot more than cacao. Xiba pointed out plants to Ismael, grabbed leaves and handed them to him. Ismael would sniff them or rub them on his skin or take a bite and then pass them on. By the time they reached me, I wasn't sure what to do with the samples and didn't want to interrupt as the group was already focused on another leaf.

We stepped into a clearing, an island of somewhat dry land surrounded by swampy forest. I tried to see the layers of the canopy through Mark Shepard's eyes. The tallest trees were the Brazil nut, their crowns spreading like giant umbrellas. They are among the longest-lived in the Amazon. Each tree pumps 250 gallons of water per day up its trunk and out through the leaves and produces fruit the size of baseballs that can weigh five pounds and plummet to the ground at velocities of fifty miles per hour, either planting themselves in the earth or killing a very unlucky person. The nuts are an important source of protein for the locals, and the shells of the fruit are used as bowls. The high amounts of selenium in Brazil nuts can help fight cancer. The trees only grow in wild and healthy forests.

**Eating the Rainforest**

Beneath them are açaí palm trees, with their hanging dreadlocks bejeweled with tiny berries, and then andiroba, and cupuaçu, and cacao, and a pharmacy and grocery store's worth of other plants, all under threat by deforestation. Xiba recently planted two thousand saplings here—bananas and pineapples and cassava.

He described a future where other farmers would come to see and learn about agroforestry, and how he harvests the natural food forest surrounding his intentional food forest. Amazon 4.0 could help make that vision a reality.

This kind of agroforestry, sustainably harvesting the natural resources of the forest, produces more revenue per acre than ranching or soybeans. And it is also one of the most effective examples of carbon sequestration. Xiba's efforts to weed, fertilize, and irrigate pull eleven tons of carbon out of the atmosphere, which is eight times more than a forest regenerating on its own, and he's certainly happier, more efficient, and more alive than any carbon-sequestering DAC machine.

Early on in my research, if someone had asked me to close my eyes and imagine a regenerative farmer, I would've pictured someone like Mark Shepard. Which is to say, a White American man who lived in the Corn Belt. That's who was depicted in news stories, documentaries, and testimonies before Congress, portrayed as a superhero standing in a field. There's no doubt that these farmers are doing important work adding diversity where monocultures once existed, but I found that there was barely any attention on farmers like Xiba who are working with existing diversity to prevent the further spread of monocultures.

The land conservation debate rages between complete protection—no extraction whatsoever—and extreme economic exploitation. Extreme economic exploitation is winning by a long shot. Seventeen percent of the forest has been clear-cut since 1970.

Amazon 4.0 is arguing for a middle ground. The thinking is that if farmers using regenerative agroforestry methods can show that their products are valued by the market, even politicians focused on the almighty GDP could be persuaded to preserve the forest. Ismael's work is just getting started and he still gets a lot of blank looks when he talks about using technology in the Amazon.

Back at Xiba's house, we sat down to enjoy a meal made only with products we'd just seen in the forest. I've been served bat, rat, and guinea pig on previous trips, so I'm always thankful when there isn't some type of unidentifiable meat. I was glad to see the bowl of purple liquid, açaí, which looked like sweet blueberry smoothie. I spooned in a giant mouthful and almost choked on the tasteless, unsweetened goo. Almost coughed and splattered purple droplets of agroforestry hope all over Ismael's face. It was like expecting orange juice but taking a big ol' gulp of milk instead.

The açaí bowls popular across the United States are sweetened to the extent that nutritionists have referred to them as "sugar bombs." Since the food industry dubbed açaí a superfood, it's been credited with everything from reversing aging and diabetes to increasing penis size. Name a modern ill or insecurity and someone somewhere was selling açaí to address it. Or as one anthropologist put it: "There are all these claims that [açaí] takes away the toxicity of living in the First World and transports you back to the healthy, natural lifestyles of those who live in the rainforest."

In the actual rainforest, I managed to swallow my first glob of açaí. I looked at the sugar on the table, and made eye

## Eating the Rainforest

contact with De Mendes. No, his look said, eat it like they eat it. Someone had climbed a few stories up a tree and harvested the hard, olive-like fruit. Then they soaked it for hours to soften it, removed the pits, pulverized the pulp, and served it to us.

Before long, my taste buds adjusted and I began to appreciate the goo, which turned out to be cold and refreshing. Açaí is a staple food for those who live along the rivers, and that truth doesn't require fancy marketing, exaggerated claims, or spoonfuls of sugar.

Xiba split open a cacao pod and passed it around like a candy bar. The exposed fruit looked like a thick cob of giant corn kernels. De Mendes took a bite and passed it to me. I bit off a few slimy seeds, sucked the sweet fruit from them, and threw the seeds over the railing into the forest.

As Xiba and De Mendes talked about the characteristics of the fruit we ate and the plants that influenced its taste, I tossed out more seeds, till I noticed that the others were saving them.

Each cacao pod produces forty seeds and every seed I spit out was a future cacao tree that could produce thirty pods, a total of twelve hundred new seeds annually. Imagine each of those seeds becoming a tree. Together they could produce 1,440,000 new trees.

Imagine the abundance of our world.

People who don't believe in magic have never planted a seed. I've never planted cacao, but I have planted a sunflower seed, and returned in a week to examine a newly emerged seedling that within a span of a few short months would become a giant flower head with more than a thousand seeds supported by a stalk as thick as a human arm. From one, comes thousands.

I'm not sure if we take such abundance for granted or if we are completely unaware of it. Maybe both.

Of course, if left to nature alone, none of those cacao seeds might become trees, but with knowledge and planning, people like Xiba can work with the abundance of nature to produce enough cacao to make a living.

As we were leaving, De Mendes reached into his small bag to retrieve three identical bars of chocolate, which he handed to Xiba. Xiba grinned and turned them in his hand to take in every detail of a cardboard wrapper featuring gallery-worthy renderings of a variety of plants and four prominent letters: X-I-B-A.

The cacao that Xiba had grown and fermented had traveled by river and road to De Mendes's chocolate factory, where stainless-steel roasters, grinders, and blenders transformed them into the bar of chocolate he proudly held.

I wanted to see where the bars were made, so when we got back, we toured De Mendes's chocolate factory, a former house a few blocks from his home. In it, he produces one ton of chocolate per month, but he wants to expand so he can support more people living in and protecting the forest.

De Mendes's nephew greeted us and gave us hairnets to wear in the factory. The main production area was a living room in which the furniture had been replaced with stainless-steel workbenches and various mixers and melters. One wall was lined unexpectedly with shelves that bowed under the weight of books protected under sheets of plastic.

Before being formed into bars, the finished chocolate is made into large bricks, dozens of which were stacked on

## Eating the Rainforest

stainless-steel shelves in what had once been a bedroom and was now a mini-Fort Knox of chocolate. Each shelf had a sticky note with a farmer's name on it.

De Mendes took a call, and his face became oddly serious as he walked out of the room. His nephew took over, and introduced us to Esmerelda, a new machine capable of ramping up their production capacity from two hundred bars to twelve hundred per day.

Bright LED bulbs reflected off the tiled floors, shiny shelves, and equipment. There wasn't a shadow. Everything was sterile and lined up and utilitarian. Except for the cacao, which was dark and lumpy. Without Xiba walking through the diverse forest in shadow and sunlight, there wouldn't be chocolate and the need for hairnets. And without this place and these people, Xiba could be forced to sell his patch of forest. Both places might not exist without the other. Together they grow.

De Mendes rejoined us, all smiles, and shared his news: Ismael, who we'd dropped off on the way back to town, had called. In a matter of a few hours, he'd conferred with his colleagues at Amazon 4.0 and had decided to work with Xiba. We celebrated by eating chocolate.

De Mendes led us into a room that stored the finished bars of chocolate, ready for shipping. I admired all the beautiful labels, which proudly proclaim the names of the people like Xiba—as well as Zeno, Sakaguchi, and others—who'd grown the cacao, their region, and the percentage of cacao in the chocolate. Each bar of chocolate represented a family legacy, a tradition, and a relationship with nature.

One bar wasn't technically chocolate at all. It was made from a tree related to cacao, cupuaçu, using a process invented by an extraordinary woman trying to save her community.

"You must visit Selma," De Mendes told me as he held up his phone with contact information and picture of a smiling woman in bright red shorts standing in a patch of vibrantly green lemongrass.

•

**Santarém, Brazil**

Selma lived in a community that was being decimated by soybeans. Ismael, continuing his work for Amazon 4.0, and I made our way there, accompanied by Davide Pompermaier, an activist helping farmers like Selma.

Celebrating their unlikely partnership, Xiba holds his finished chocolate bars, professionally produced and beautifully packaged by De Mendes. KELSEY TIMMERMAN

## Eating the Rainforest

On the way to her home south of Santarém, which is five hundred miles up the Amazon River Basin from Belém, we passed giant fields of corn and soybeans and every so often a system of grain bins and elevators and John Deere dealerships. It was eerie how similar this view from the window of a truck in the middle of the Amazon was to the one from the window of a truck in the middle of Indiana.

Agrochemical companies strive to develop seeds and chemicals that work across vast areas, the larger the better. They seek to provide a select few solutions that answer all problems, and, to them, every living thing that doesn't spring from their seed is a problem. Davide's organization, the Health and Happiness Project, does the opposite. Focusing on water and sanitation, family health, and preventative care, the group works in remote areas, supporting diversity and sustainable development, always recognizing that each of the many communities it's involved with is different and requires its own solution. They also develop cultural experiences, including a radio station and circus.

"We help communities find a way to live with a different world," Davide said as we traveled down BR163, known as the Soybean Highway. Davide estimated that three hundred to four hundred truckloads of soy headed down it to the port in Santarém each day.

A sign pointed ahead to the town of Rurópolis, where the Pan-American Highway crosses the Soybean Highway. The aptly named town sits at the intersection of the conversion of rural areas into cities.

How does this conversion happen? First, rural people are forced from the land. Then loggers clear-cut the forest and sell off the lumber. Burning sends the forest's remaining nutrients into the ground, but they're sucked up quickly by whatever crops are then planted, and before long, farmers have no choice but to manufacture life from nothing, turning to GMO seeds and chemical fertilizers produced from other mined resources, including natural gas and coal. The resources and wealth and people from the rural communities pass through Rurópolis and are lost to the cities.

To feed cattle and the growing demand from China, in the 1990s and 2000s farmers in this area began to clear thousands of acres at a time and then plant huge swaths of land with soybeans. A port was constructed in nearby Santarém, although someone new to town arriving by ferry would be excused for thinking the city was named Cargill. The name is plastered in giant lettering at the port. The American company constructed a processing facility there in 2003, and a "soya rush" began. Cargill recruited farmers to expand into soybeans and helped finance the expansion. Since the 1970s, 17 percent of the Amazon has been lost to croplands and pastures. And globally, agriculture expansion accounts for almost 90 percent of deforestation. "Soybeans changed this town," said Davide.

When we arrived at her place, Selma, fifty-two, greeted us with youthful energy and passion. She immediately led us to some picnic tables covered with tablecloths and a spread of cake, bananas, truffles, coffee, and lemongrass tea. As we dug in, we quickly learned that everything wasn't as it seemed. Actually, the bananas were legit, but the truffles, which looked like chocolate, were made from cupuaçu. Selma referred to them as *cupulate*, a play on "chocolate." She had perfected the processing of cupuaçu into a chocolate-like delicacy and taught it to De Mendes, who used the recipe with her permission in the cupulate bars we'd seen at his factory. Also, the coffee wasn't coffee, but roasted açaí seeds steeped in hot water. There was no lack of creativity or effort in our snack. Selma was trying to showcase how versatile her community's crop of cupuaçu could be, hoping Ismael could help bring it to market.

Selma's family moved to this area when she was fourteen in search of land to farm. Her father had an axe and started cutting. They built a simple log home. She remembers that at first they couldn't sleep, afraid of the howler monkeys that would roar and bellow, claiming their treetop territory in the forest before sunrise. In time, they adapted.

"We learned to live together with the animals on the edge of a stream," Selma said, "totally in nature. It was very good."

But there were no schools, so her family sent her to Santarém to study, and she eventually met a guy, had kids, and dropped out of school.

When she talked about where she lived then and where she lived now, she always referenced how many kilometers it was from Santarém. Why is it that we measure how far we are from "civilization," and not how close we are to nature?

She and her new husband moved to her family's farm, 135 kilometers from Santarém, but eventually they bought this farm at 140 kilometers and planted black pepper, cassava, wheat, rice, and beans. They lived off what they and their neighbors grew. They shared food and they shared work.

"We lived on our own support without salary," she said. "For me, this is paradise."

As the roads opened up, the loggers came, and then the soybeans. Pressure mounted. Her husband cut the forest to make more fields.

"Since I was little I was always worried about anything that destroyed the environment," she said. "When my husband was going to the fields to do a big felling, I was at home in despair. I questioned the reason."

She thought about the immensity of the forest and felt its loss. "The forest was crushed by the weight of huge tractors, felling everything to plant soybeans. Animals and people had to leave to who knows where. I was devastated to see it all."

Selma decided she had to do something about it. So for every tree her family cut down on their farm, they began to plant others—native trees such as açaí, cacao, and andiroba.

Life on the farm was good, but she moved with her three daughters to Santarém so they could be educated. They lived in a one-room apartment and Selma got a job as a maid and as a janitor at the kids' school. Once they'd graduated, she finished high school herself.

When Selma moved back to kilometer 140, she grieved the changes she saw in her community. People were selling their trees and then their land, getting jobs with large companies cutting, expanding, building. Selma wanted to fight back, so she got involved in the union of rural workers. After becoming frustrated that so many of the union's efforts were focused on men, she founded AMABELA. "Ama" refers to loving and "bela" to women. Her association of loving rural women workers promoted the idea that they have a "right to freedom, to fight for your dreams, to build and lead your life according to what you believe."

She and other AMABELA members attended a training program on territorial rights. "I learned a lot," she said. "I began to see that the big businesses, whether the soybeans, the mining companies, and so on, were only able to grow by destroying, leaving our people in poverty."

At another training she learned about agroecology—applying ecological concepts to farming—agroforestry, and the critical role women played.

"Just as we wanted to be respected and valued as women," she said, "we should also value Mother Earth's feminine spirit and all that she can offer us. I believe that when we work with love, we move forward. We have to take care of our Mother Earth with

A logging operation on the banks of the Amazon River. Once the trees are removed and the land burned, the soil where a diverse rainforest once stood is planted with soybeans, resembling a farm in the American Midwest. KELSEY TIMMERMAN

great love, as she will repay us by giving us everything we need to support our families."

Selma led us beneath the canopy of açaí trees, our shoulders rubbing against cacao and cupuaçu. Ugly chickens with long, featherless red necks scraped at the foliage on the ground.

Soon, we were back at the picnic tables, where eight members of AMABELA, ranging in age from their twenties to their fifties, joined us. Ismael gave a presentation on his laptop about Amazon 4.0 and the resources they make available to communities they work with. He told the group that he really liked their açaí truffles, and that he was impressed by the De Mendes cupuaçu bars. And he said there might be some opportunities for Amazon 4.0 to assist them in developing cupuaçu. Selma wiped away tears.

I looked around and noticed the fresh white paint on the house. Brown paper bags of cupuaçu were lined up perfectly for everyone to sample. The sandy driveway and courtyard had lines raked all the same direction, only disturbed by the tires of our vehicle. There was barely a leaf on the ground. Selma had put a lot of effort into making a good impression.

I joined Davide in the courtyard away from the group.

"She's nervous because she needs help," he said. "Without technical assistance, more and more people are selling."

Cupuaçu has potential, but it will only work if it's scaled up, if many small farmers like Selma work together, sharing the cost of production and of getting their product to market. Unfortunately, AMABELA has shrunk from eighty-two members to just forty-five. Each loss weakens the collective. Davide said the membership was spread out over a large area and doesn't seem to have a cohesive vision.

Later, I asked Selma, "When was the last time someone made an offer to buy your land?"

"Two weeks ago," she said without hesitation. They offered her $35,000 for her hundred-acre farm. That's about thirty times less than the cost of farmland in Indiana, but a nice sum in Pará.

"And they called everyone around you?"

"Yes," she said. "My neighbors sold their farm two weeks ago." She told us that as neighbors are replaced by soybeans, her once-full church is becoming empty.

"Where will they go?"

"To Santarém, of course," she said. "But the people who move from here to Santarém don't have much to do there. They are rural farmers."

"It's like this where I live, too," I told her. "My home is surrounded by corn and soybeans. People leave the countryside to go to the city, our rural communities are getting smaller and smaller, and the farms are getting bigger and bigger."

I wanted to say something encouraging, but it all felt phony, so I just smiled and we hugged.

Back in the truck, Ismael said he wanted to see what he could do. If they could help Selma hold the line against soybeans, it would be a big win.

That seemed hopeful. And then I asked the question I didn't really want to ask: "Do you think the community really has a chance?"

Ismael hesitated and said, "Ultimately, no. Because they literally face a monster."

Later, Selma sent me a poem she'd written:

> Beautiful land never fed by pesticides,
>
> The great beauty fades away.
>
> And starting with roots,

Selma in her AMABELA shirt giving a tour of her farm. She hopes that her chocolate-like creation from cupuaçu can provide enough profit to prevent the members of her collective from selling their land to industrial farmers. KELSEY TIMMERMAN

**Eating the Rainforest**

*Our people have already been condemned!*

*Swallowed by the ambition of soybean corporations, by millions of land grabbers,*

*Who aim only at profit, agribusiness and money.*

*On the BR 163 there is crying and gnashing of teeth,*

*Trucks crushing everything, even people.*

*But we offer a call to life.*

*With the resistance of rural workers, Indigenous people, and outsiders.*

*They still take care of the river, embracing the beautiful enchanted land, our beloved land.*

- 

**Tapajós River, Brazil**

The generator sputtered and then hummed. The PowerPoint presentation in the middle of the Amazon began. The presenter, Dr. Raquel Tupinambá, had her hair pulled up into a topknot. The projector sat slightly askew to account for the crooked screen—a rectangular piece of cloth that hung on mats woven from local vegetation.

Raquel was one of the youngest leaders of the Tupinambá people, who live along the Tapajós River a few miles upstream from its confluence with the Amazon, as they had for centuries, likely millennia. Her community wasn't that far as the toucan flies from Selma's, across the river, but it was worlds away. On this side there were no roads, no soybean highways—but the threat of logging, farming, mining, deforestation, and the loss of her culture were immediate and real.

Raquel had a feather hanging from her right ear. She also had a PhD in anthropology and a master's in biology. This wasn't her first PowerPoint. She clicked through her slides, presenting to Ismael and one of the founders of the Health and Happiness Project, Eugenio Scannavino Netto, who sat in plastic chairs near the front of the open-walled building with a thatched roof and a concrete floor. Her fellow community members sat on tree-trunk benches as she explained the threats to her community and their efforts to fight back.

Raquel told us that eight million people lived in the Amazon before the conquistadors arrived, surviving on an abundance of native plants—138 species. She clicked to an image of a house next to a river surrounded by a cleared field with no forest in sight. The title of the slide was "Amazon versus Humans."

"Traditional knowledge and the entire legacy of family farming are losing their value to the conventional agricultural model," she said.

Conventional agriculture, logging, and mining are based on the idea that the only way to make money is by destroying the forest, but the Tupinambá community was challenging that. They were determined to prove to the Brazilian government that the forest—their home—was not a valueless void, but quite the opposite. It was filled with richness, everything they had ever needed. Still, Raquel was worried that her people might begin to believe the government line that they were poor and needed to join the outside world. She started to talk to her community members as much as us.

## Eating the Rainforest

Raquel explained how rich their culture and homelands are and then continued, "It is necessary that we break with this idea of poverty that is internalized in the imagery created about the Amazon and Indigenous peoples. We continue to resist with our ways of life, guaranteeing the good life and protection of our territories."

I thought back to the Arhuaco and how at first glance, I thought they were poor. I suppose they were, if poverty means a lack of material things. But I have the feeling the Arhuaco and the Tupinambá would look at my culture's lack of knowledge about and connection with nature, not to mention our weaker family and community ties, and would agree to use the same term to describe the way I live.

Her audience—even the kids—sat in quiet support of all that she had to say. Occasionally they looked to Ismael to judge his reaction. Eugenio gave me a nod that said, "She's impressive, no?"

I nodded yes. Raquel, a young Indigenous leader, activist, and scientist, was using the experience, knowledge, and skills she'd gained from outside her community to organize and fight to preserve it.

Raquel ended her talk. When someone turned the generator off, its thumping was replaced by the distant whine of a chainsaw.

Eugenio stood to talk. The cadence of his speech was song-like and even though I could see that the subject wasn't funny, the effect was like a standup routine. Talk. Talk. Everyone laughed. Talk. Talk. Everyone laughed. He wore board shorts, a baggy white shirt, and Croc sandals, one of which was missing a strap. He's a medical doctor, a man of science, but has used humor,

clowns, and puppetry to help build connections with this community and was now using humor to reinforce them.

Then it was Ismael's turn. I could tell he was tired. His eyes were bloodshot. I had been traveling with him for almost a week, and had seen how he had to always be "on," listening, speaking, trying to figure out if Amazon 4.0 could help a community or not. He had committed much of his life to the rainforest, but he really wasn't built for it. He's a big guy, maybe six-four, which meant he had ducked under a lot of branches, sunk into a lot of mud, and sweated through a lot of shirts. It was an especially hot day, but he was as passionate as ever.

Dr. Raquel Tupinambá, a member of the Tupinambá people, who are indigenous to this part of Amazonian Brazil, delivers a PowerPoint presentation on how her community is threatened by extraction and how they are trying to preserve their lands.
KELSEY TIMMERMAN

## Eating the Rainforest

He talked about the potential of cacao and Selma's cupuaçu creations. His team was working to design a mobile processing center to turn both into candy bars. I thought of how even if Selma lost the fight against the land grabbers in her own community, her efforts would help others, like this one, win.

I couldn't have predicted that my regenerative journey would bring me to this place to meet these people. Just like with Celestine and Dalmas, there was no talk of cover crops or chemicals or new farming technologies—and no question of whether or how agriculture should be regenerative. It just was. The forest and the Tupinambá culture are regenerative and have always had to be in order to survive thousands of years in the Amazon. Although outside processes and machines could help the Tupinambá resist the political and economic forces they face, it all felt so small. Even with boatloads of the latest and greatest modern gadgets and technologies fit for the rainforest, I wasn't confident that humanity could engineer our way to a regenerative future.

But it's not just about technologies and methods. It's about relationships. Eugenio had a long-standing relationship with the Tupinambá; Ismael was building one and sharing about Selma, who had discovered a new relationship with the cupuaçu plant. I was realizing that if agriculture was going to be truly regenerative, our ideas may not save us, but our relationships might.

The Tupinambá's products, which they sold in Santarém, were displayed on a handcrafted table topped with banana leaves. There was cassava wine, an ancestral cassava sauce, flour, and cupuaçu jellies bearing the brand name AMPRAVAT, which represented the collective efforts of thirty families. Raquel and her sister, who has a degree in agroecology, had tested forty different varieties of manioc to find the best one for a drink they'd produced.

I was impressed with these efforts and of course with Selma's, too. These creative and dedicated people were trying so hard. But it all left me with an uncomfortable feeling in the pit of my stomach. Could selling such products save them, save the forest? I had to wonder. And what would be the unintended consequences of any success? What if cupulate, for example, became really profitable, much more than any other product? Would Tupinambá farmers choose to grow more of it, undermining the overall biodiversity of the forest?

It makes sense that these questions would occur to me because I come from a world where every consumer desire is met. I'm encultured in growth; I'm the child of entrepreneurs. How many times did I hear my dad say, "If you're not growing, you're dying"? For the family's construction business, that meant using more wood and steel and oil and fuel to make more money.

But this mindset I was born into isn't hardwired into our DNA. If any group of people can dip their toes into the river of capitalism and not get washed away on the river of growth, it's Indigenous and traditional societies like the Tupinambá. Their deep roots anchor them to the land and generations of living with nature. The Arhuaco in Colombia told their non-Indigenous coffee farmer partners that they didn't want to fix the roads or expand their production. Cultures built on reciprocity and gratitude for nature have a better understanding of something Western societies struggle with: when enough is enough.

As Raquel stated, the Amazon rainforest long supported populations of millions—even densely populated cities—mainly through large-scale, diverse agroforestry. That balance was thrown off by the death and disease brought by European settlers starting in the fifteenth century, and also the unsound introduction of monocultures.

The first monoculture in the Amazon was rubber trees planted by the Ford Motor Company. In 1928 Henry Ford founded Fordlandia, a community neighboring the Tupinambá that he envisioned as a pastoral utopia that would provide a cheaper source of rubber for his company. He claimed, "We are not going to South America to make money, but to help develop that wonderful and fertile land." He believed that industrial capitalism could provide opportunities for the local people. He paid famously high wages.

However, the abundant life of the jungle quickly consumed his trees with foliage and disease. The crop was a failure, as were the class divisions built into Fordlandia, which provided American workers with the best view of the river and running water as the Brazilian workers lived in houses described as "midget hells where one lies awake and sweats the first half of the night, and frequently between midnight and dawn undergoes a fierce siege of heat-provoking nightmares."

The experiment of Fordlandia did not, in fact, help the "wonderful and fertile land." It did quite the opposite, his other big farming ideas suffering as unequivocal a death as rubber. Greg Grandin writes about its failure in *Fordlandia: The Rise and Fall of Henry Ford's Forgotten Jungle City*: "In the Tapajós valley, three prominent elements of Ford's vision—lumber, which he hoped to profit from while at the same time finding ways to conserve nature; roads, which he believed would knit small towns together and create sustainable markets; and soybeans, in which he invested millions, hoping that the industrial crop would revive rural life—have become the primary agents of the Amazon's ruin, not just of its flora and fauna but of many of its communities."

Back in the twenty-first century, Ismael, contending with these same issues, was done with his presentation, so we took a walking tour of the community. We passed a naked three-year-old holding a wooden plank. He pretended to pull something from the end and then began making chainsaw noises. Another boy beside him, perhaps his brother, held the bar of a real chainsaw and cut down a pretend tree.

Eugenio told me that not everyone is able to stay in the community or come back, like Raquel. Many of the kids leave to go to secondary school, living in slums at the edges of cities in order to learn how to exist in the encroaching culture.

"They don't want to go," Eugenio said, "but they do go."

A solar array installed with support from the Health and Happiness Project sat behind a community building housing the processing equipment Raquel's group used to make their products. An interior wall was lined with a bank of thirty car batteries, storing the solar energy to run the equipment and freezers. At another location, a building that housed a well had walls stained from black exhaust from a no-longer-there diesel generator. Now the building housed a bank of batteries storing the solar energy to pump water.

I could only imagine how diesel arrived. By boat? Carried up the beach? However it got here, it wasn't as reliable as the sun.

The midday sun beat us up. Ismael looked progressively like he'd been punched. The hammocks surrounding houses were occupied.

I applied a constant slick of sunscreen and rotated the bill of my hat to maximize protection.

Next up was a tree nursery. Like Ismael and me, the seedlings were sensitive to the sun and needed the shade provided by nylon cloth hanging limp in the breezeless afternoon. The nursery was a copy of the one at the Experimental Active Forest Centre (CEFA), downriver from Raquel's community, which the Health and Happiness Project had established in 2016. CEFA leads a host of efforts to teach community members to incorporate their crops into planned agroforestry systems.

"We are changing the model," Eugenio told me. He's particularly interested in encouraging farmers to replace manioc, an annual plant, with perennials and tree crops such as cacao and cupuaçu.

Products the Tupinambá produce and sell include cassava wine, an ancestral cassava sauce, flour, and cupuaçu jellies. Their profits demonstrate that the forest can provide for people without being torn down. KELSEY TIMMERMAN

The Health and Happiness Project worked with an agronomist to plan the CEFA nursery with the goal of it serving as an example of agroforestry that the surrounding communities could adopt. The agronomist chose the location because the soil was sandy and part of the area was a savanna where only an invasive grass would grow. If they could regenerate this land, other communities would be able to regenerate theirs.

This is a common theme I've noticed among people with a regenerative mindset. The worse off the land and the more undesirable it is to others, the more the regenerators seem to be drawn to it. I would see this time and again during my travels, as they'd start with soil that had been farmed to death or land deemed worthless. At first it seemed like madness or cockiness, but the more I came into contact with it, I saw it more as a faith in nature. Plant some trees, give nature a little kick-start, and then let the sun, the rain, microorganisms, and fungi take over.

It was a success at CEFA. After two years, the plants were thriving and a river that had dried up long since began to run again. When you witness a resurrection, you believe. Community members started to trust in the agroforestry system.

And Eugenio believed in the Tupinambá.

After a long, steamy day, we made our way back to the boat we'd arrived on.

Eugenio paused before boarding. "Would you like to swim?" he asked, slipping out of his board shorts and walking into the water in his Speedo.

"Absolutely," I said. He didn't have to ask twice.

## Eating the Rainforest

The water was flat calm. I dove below the surface and listened to the hum of the Amazon.

I exhaled as I broke the surface. It was only then I thought about piranha, alligators, and anaconda, but I'd learned there were much scarier things in the rainforest.

There were bad guys. Just as in *RoboCop*, the bad guys were corporations and people seeking to profit from the suffering of others. Or maybe it wasn't that simple. My Grandpa Timmerman loved Ford. Maybe his first truck contained rubber made from latex shipped down this river. Maybe Cargill bought the five acres of soybeans produced from the field I played in as a kid, helping make my happy memories possible. Maybe if I could snap my fingers and don my rainforest warrior RoboCop suit, I wouldn't know who to fight. The heroes of this story are people like Selma, who fight against long odds; Xiba, who hopes to educate his neighbors about growing cacao; and Raquel, who stands and speaks and carries on the traditions of her people. I don't fight for them. They fight for us, and a whole world that benefits from a healthy Amazon rainforest.

As Eugenio wrote in an op-ed in a São Paulo paper, linking our disregard of nature to the COVID pandemic: "We could observe [nature], respect her and learn to interact harmoniously... with her wisdom and generosity... as most traditional populations do. For this reason, I sometimes believe that the future will be Indigenous."

Maybe our role isn't to come out of nowhere and try to save the rainforest, but instead to support those who can.

# Part II: Rethink

# Chapter 6
# Hawaiian Pride B 4 Pesticides

# Kaua'i

**Regenerative Agriculture Rejects Industrial Agriculture**

Malia Chun stood on the side of the road in Waimea on Kaua'i's west side, holding a sign.

She stood with her two daughters.

She stood for her two daughters.

Waimea is small, fewer than two thousand people. It has a few fish taco and shaved ice places catering to tourists on their way to Waimea Canyon, but there aren't any hotels, resorts, or shiny condos. For the most part, it's locals trying to make a life.

Sugar plantations left empty, hulking factories on the edge of town. The buildings' bones rusted, the salt air trying to erase their existence and turn them into soil, something useful. Some locals live on the property, which is still owned by the sugar company. The long-term, rent-free leases on the plantation houses are for Native Hawaiians, who according to the US government must be at least 50 percent "part of the blood of the races inhabiting the Hawaiian Islands previous to 1778."

Malia is thankful for her home there, which was passed on to her by an uncle. Many Hawaiian residents struggle to afford housing and there's no way, as a single parent, she could manage to buy a place. Still, she thinks of communities like hers as

Malia Chun, a mother of two, stood up proudly against the agrochemical companies harming her community. MIKE COOTS / EARTHJUSTICE

holding pens, like reservations, and views the blood quantum requirement, which hasn't been changed in a hundred years, as a way to slowly bleed away Native culture until there are no Hawaiians left, further removing them from the land.

From her yard, looking north toward the mountains, there's open land as flat as a field in Indiana. There's another similarity, too: what's in the field is corn. This isn't some tropical variety. It's field corn, the same as what's surrounding my home. In fact, the seeds planted in places like Indiana, Nebraska, and Iowa were likely developed on Kaua'i, which has the advantage of year-round growing. Though we like to assign heritage to place of origin, it turns out that "Indiana corn" may actually be Hawaiian corn (or Puerto Rican corn, or even corn from Chile or Brazil).

Within a few weeks of Malia and her girls moving to their inherited home, they began to wake up in the middle of the night with their eyes burning. They had trouble breathing. They all developed asthma. The doctor said it was likely something in their environment.

"What the hell?" thought Malia, who was in her thirties and had never had asthma. "Why am I having to use an inhaler?"

She noticed that the corn was sprayed after dark.

"And then I'm like, 'Okay, what are they spraying? Why are they spraying at night while we're sleeping?'"

Answers weren't easy to come by, but she discovered that Syngenta, a Switzerland-based agrochemical multinational corporation, conducted the spraying. Her community is completely surrounded by chemical companies.

## Hawaiian Pride B 4 Pesticides

The more Malia learned, the more concerned she became. And she wasn't alone. Residents on the east side of Waimea had just sued DuPont-Pioneer for damages caused by spray drifting onto their homes. Pioneer had converted sugarcane land to seed corn and GMO research in the '90s. The first time one resident smelled the chemicals, he thought his neighbor's house was burning down. After some complaints, the smell didn't stop; it changed. Bubble gum. That's how one resident described it.

Pioneer's research center and fields sit on a bluff above the Waimea River and the neighborhood. The winds carried red soil and who knew what else into the river and down onto their houses, so residents closed their windows and installed central air. And they wondered if their rashes, cancer, asthma, and nosebleeds had anything to do with the bubble gum wind or polluted water.

The lawsuit was still unresolved when Malia started asking questions in 2012.

She was aware of what had happened at Waimea Middle School a few years earlier, in 2008. Students had been playing capture the flag on the playground when they also started to smell something weird. They became nauseous and dizzy and ran inside, where some of them began vomiting. This event and others like it through the years sent twenty kids to the ER. A teacher measured what was in the air and found traces of the endocrine-disrupting chemicals atrazine (which was banned in Syngenta's home country of Switzerland and across Europe in 2003), metolachlor, and chlorpyrifos. Syngenta attributed the incidents to hysteria and to the nearby removal of a noxious weed.

Seeds, mainly corn, are the number-one agricultural product of Hawai'i, at its peak contributing $241.6 million to the economy and perhaps more importantly to the residents of Waimea, providing fourteen hundred jobs. To fight for clean air in your neighborhood was to fight against your neighbors' employer. So the residents of Waimea and neighboring Kekaha largely kept their opinions of the seed companies to themselves.

"Nobody was doing anything about it," Malia said. "When it started, I felt like I was alone."

Malia began to go door-to-door talking to other moms.

"You guys getting asthmatic symptoms?" Malia would ask.

"Yeah," they'd say.

"Are your kids' eyes burning?" she'd ask.

"Oh... yeah," they'd say.

But nobody wanted to make the connection to the fields. They thought she was crazy. Or they couldn't ask questions because they had an uncle who worked there. But she saw it differently. Here you have one of the largest Native Hawaiian populations, a rural community, and they're encircled by Syngenta, Bayer, DuPont-Pioneer, and Dow Chemical. The residents were under siege by land, air, and water. She saw it as cultural genocide.

"And the thing is," Malia said, "I'd just feel like a complete hypocrite if I didn't fight for the things that I was trying to instill in my kids."

So when knocking on doors and talking to moms in private didn't get results, she grabbed a poster board and some markers. In a

The remnants of a mill on an abandoned sugar plantation on Kaua'i. Much of the land that once grew sugarcane now produces up to three harvests of seed corn a year. DENISE ANKRUM / SHUTTERSTOCK

Josh Mori tending the aquaponic garden at the local high school. Josh was one of the first to join Malia's protest against the agrochemical companies. KELSEY TIMMERMAN

## Hawaiian Pride B 4 Pesticides

small town where everybody knows everybody and feather-ruffling is discouraged, she stood with her daughters and held her sign for all to read: "Hawaiian Pride B 4 Pesticides."

And it turned out Malia wasn't alone.

A white Toyota truck stopped and a local guy wearing only surf shorts jumped out with his pit bull and said, "I've been waiting for you!"

Josh Mori's head was clean shaven, and he was built like a football player. Because Josh Mori *was* a football player. He played cornerback at a college in Oregon before returning to Hawai'i, where he assisted with a traditional Hawaiian fish pond alongside his mentor, Walter Ritte, known as "Uncle Walter," who has been fighting for Hawaiian rights, culture, land, water, and environment since the 1970s.

Now Josh was the athletic trainer for the baseball and football teams at Waimea High School, and a lot of the dads of the kids he coached worked for Pioneer and Syngenta. That made it awkward, but he wanted to do something. "Do you have a sign for me?" he asked.

Malia handed Josh a sign. They talked for hours. They learned that they had each participated in a recent protest in Poipu, a resort town fifteen miles to the east. There's even a picture of them side by side, unaware of the future ally and friend next to them. Along with thousands of others that day, they'd marched against the agrochemical companies and for Kaua'i County Council Bill 2491, which would require the companies to disclose what was being sprayed when, and limited how close the spraying could be to homes and schools.

Once the two met, "from that point on we haven't stopped," Malia said.

Ultimately Malia and Josh would join a fight against some of the largest corporations in the world. It would change their lives, the lives of countless others, and alter the landscape of Kaua'i.

●

I came to Kaua'i to learn how the agricultural research on the island impacted what seeds were planted and what sprays were applied next to my house in Indiana and around the world. But I also wanted to see for myself how a small group of activists, Hawaiian culture, and regenerative farming practices played a critical role in the fight for the island's food security and sovereignty. I was seeing that changing agriculture isn't just about changing how we farm in the field. It's also about how we protest, organize, and address injustice. What I saw activists like Selma, Raquel, Dalmas, and Celestine doing in Brazil and Kenya sometimes felt dangerous and hopeless, but here in Hawai'i a native community had risen up and against long odds, pushed against industrial agriculture and toward regenerative practices.

●

"Let's plan on doing the toxic tour Friday morning," Gary Hooser texted me soon after I arrived on Kaua'i.

Gary was the Kaua'i County Council member who, along with a colleague, introduced Bill 2491 in 2013. Others, like Malia and Josh, had lit the match, but essentially Gary had set the metaphorical stick of dynamite on the council's desk.

## Hawaiian Pride B 4 Pesticides

He picked me up in his silver Toyota truck and we headed toward Lihue, the county seat.

Gary has lived on several Hawaiian islands but has been on Kaua'i since 1980. His resume is full of character, including everything from state senator to pedicab pedaler, and his look is a caricature artist's dream—wavy hair, bushy gray eyebrows, and a sparkling politician's smile. I can barely draw a stick figure, but I bet I could draw Gary.

It was a weekday, but because of the pandemic, the parking lot of the historic county building held more chickens than cars. The Hawaiian print on Gary's mask clashed with the one on his shirt. He poked his head in the door and asked if he could show me around.

Of course, the man there knew Gary. Everyone knew Gary. There was his time on the county council, but because of his weekly column in the paper, Bernie Sanders bumper sticker next to his own (it read: "Hooser: Local Roots. Democratic values. People first."), he's hard to miss. He was a state senator for eight years, ran and lost a race for lieutenant governor, and was appointed by the governor to be director of environmental quality for the state. Gary had opinions and let you know them, and he worked through his nonprofit, HAPA (Hawai'i Alliance for Progressive Action), to turn those opinions into policy.

The council chambers were smaller than I'd expected. Desks for the council members took up half the room, and polished wooden seats for an audience the rest. Gary showed me a framed photo of the council that passed the bill in 2013. He had brown hair when it was taken.

Most people associate Hawai'i with lush cliffs plunging into the deep blue sea, but the landscape also features acres upon acres of monocropped corn. KELSEY TIMMERMAN

As Gary described what had transpired there, he held out both hands and vibrated them, trying to communicate energy in the packed room and the chants heard from the crowd of thousands of supporters and protestors outside.

"I've been involved in a lot of stuff," Gary said, "but this was a life-changer for me."

•

Gary hadn't seen this fight coming. At first he'd get only the occasional community member complaining about the spraying. And then people from all walks of life and all over the island started to approach him. People who lived in his north-side neighborhood, filled with transplants, wealthy tourists, celebrities, and billionaires, were concerned that the seed companies were going to start planting near their homes. A surfer on the south shore left Gary a voicemail. Through coughs and chokes, he said he'd been overcome by a smell at Polihale State Park and saw the spraying.

And then one day, Gary got an urgent message from one of his son's friends, Sol Kahn: "Uncle, we need your help!"

Sol was a surfer, model, and hunter. He thought the anti-GMO fight was something led by crazy hippies who'd moved to the island and never left. He figured the seed companies he'd drive by on the way to the beach were making seeds for the world and were "cool mainland businesses" that brought in jobs. But then he started to hear about dead sea urchins washing up, and turtles and pigs with tumors. He began to research industrial agriculture and concluded that all of the negative impacts he'd seen were linked to the chemicals the companies were applying to their fields of GMO corn.

## Hawaiian Pride B 4 Pesticides

Sol invited a few of his friends to meet with Gary in his living room. Only two showed, including Fern Holland, a young activist born and raised on the island who had studied environmental science and sailed with anti-whaling ships in the Pacific. They wanted the agrochemical companies gone.

"You know I just can't kick them out?" Gary told them. "Politically, I can't do it. Legally, I can't do it… but let's talk about what I can do."

Maybe he could get the companies to disclose what they were spraying and maybe he could get some other basic questions answered.

Gary arranged a meeting with representatives from the companies and put on what he called his "Columbo routine." Columbo was an underestimated 1970s TV detective who would fumble around an investigation before getting to a cutting question with his catchphrase: "Just one more thing."

Gary had started with what he thought was the easiest of his "maybes": Maybe he could get a list of the chemicals they were spraying? Maybe the companies could tell him how much they used restricted-use pesticides—the ones the EPA deemed more dangerous that are not available to the general public?

No, they told him. That was proprietary information, but they assured him that "it's safe."

Gary had another question, just one more thing. Why is it that the employees who applied the chemicals wore hazmat suits?

And then, as if it had anything in the slightest to do with the issue at hand, a company representative told him that they were "feeding the world."

Gary knew that humans don't directly eat field corn; after all, the boxes of corn shipped off island were stamped "Not For Human Consumption." A third of the corn went to feeding livestock, 40 percent to ethanol production, and what's left we eat, but only after a lot of processing to turn it into things like corn syrup, cereals, and tortilla chips.

The meeting got him nowhere. Dejected that the Columbo routine hadn't worked, Gary drafted the bill that would become known as 2491, mandating the maybes.

"And when I introduced it," Gary said from the empty council chamber, turning from the pictures of the council members, "you would have thought the world was going to come to an end."

•

The police estimated that there were well over a thousand protestors at the 2013 march in Poipu, three months later. Fern, who was born at the end of an aerial test strip where the US military had tested Agent Orange on Kaua'i, paid for the port-o-potties herself. That's just one of the hoops she had to jump through to get the approval for the mile-and-a-half march, which required a parade permit.

While Gary was working on the bill, Fern was organizing. She grew up on the island. She can be fancy and hang with her north-side movie star friends, including James Bond actor Pierce Brosnan, but she also might spend days on the west side, working outside in the sun and rain as an environmental consultant. She's what the locals call a Tita, meaning she's tough and doesn't take shit.

There was an anti-GMO movement sweeping the islands that she and her fellow activists tapped into. Fern had snuck into her stepdad's office, burning through the ink on his printer for flyers, and a local tattoo artist also let her use his printer and paper cutter.

Their efforts were so successful and so many people turned up that Fern had trouble getting to the start of the march. Poipu is known for being dry, and yet on that day the sky opened up and rain, viewed as a blessing from the Hawaiian ancestors, bucketed down. She ran in her squeaking wet flip-flops to the start of the march.

Fern, who had worried that no one would show, heard the marchers before she saw them. She ran around the bend and no longer felt the pain on her blistering feet.

"I will never forget the feeling," she said. "I see this line of thousands of people marching toward me. I felt like I was flying."

There were activists, residents from Gary's district, and people from other islands who fought against the same companies, but most of all Fern was lifted by the number of locals and Native Hawaiians. When she reached them, she turned around, and she's not sure why, but she raised her fist into the air and shouted, "Charge!"

The eyes of police officers controlling traffic at the approaching roundabout widened. But Fern had it under control. She ran ahead to the officer she had worked with to organize the permits.

"Lieutenant Rosa, what can I do for you, sir?" she said, perhaps too intensely.

"You've got to get these people in one lane," he responded.

Fern ran back to the marchers and yelled, "One fucking lane!" and started pushing people onto one side of the street. They complied.

An old woman holding a poster of her deceased daughter had blown out her flip-flops. Fern slipped hers off and gave them to her before dashing back to the officers at the roundabout.

An impressed Lieutenant Rosa introduced her to the police chief. The chief held an umbrella, keeping his perfectly pressed uniform dry. Fern was soaked, water cascading down her in rivers. She stuck her hand beneath the chief's umbrella.

"Nice to meet you," she said, and then asked if they would please step out of the way.

The police watched as the marchers passed. Some flew Hawaiian flags from long bamboo poles. Some wore hazmat suits.

"Hawai'i is a chemical test site!" Fern yelled. "You are chemical test subjects!"

•

As we drove along on our toxic tour, Gary couldn't remember exactly where the road to Poison Valley was. It was somewhere in these Waimea Mountain foothills, and he was determined to find it.

We stopped on a dirt road to get our bearings. I've never felt so awkward standing among fields of corn. They stretched a mile to the mountains to the north; to the south was a mile of corn all the way to the ocean. In some fields, corn was nearing harvest, and in others it had just pushed through bare soil. And it felt like we were somewhere we shouldn't be.

I noticed that the type of trucks we passed on the road had shifted from the Toyota "surfer's trucks" like Gary's to much larger Fords and Chevys—farmer's trucks. But here there were no trucks on the road. Just us.

Gary had driven past a sign that read: "Stop: Unauthorized Entry Strictly Prohibited. Violators will be prosecuted." It didn't state what authority would be doing the prosecuting. Maybe it was the US Navy's Pacific Missile Range Facility (PMRF), the "world's largest multi-environmental range," able to launch missiles to the ocean's surface, underwater, into the air, and into space. Hypersonic weapons, which a recent US president referred to as "super-duper" missiles—they fly five times faster than the speed

This march in Lihue, Fern's in Poipu, and others drew thousands of protesters demanding to know which chemicals were being sprayed—and when—near their homes and schools. DYLAN HOOSER

of sound and are capable of hitting a target thousands of miles away in minutes—have been launched here, too.

Corn makes for good neighbors to a high-security military base, at least if the living quarters are far enough from the spraying, which they seemed to be. We couldn't see any buildings or any people. But I had a feeling they could see us.

And if the military didn't know we were there, whoever was piloting the high-clearance sprayer floating over the field of corn certainly did.

The road ended at a serious fence and a sign that looked like it really meant business.

"Restricted Magazine Area," Gary read the sign.

"What the hell does that mean?" I asked.

"Bombs. Explosives. Ammo," Gary said. "I'm surprised they even let us back here."

We gave up on finding Poison Valley and turned around.

I was relieved that Gary had finally gotten the point. As we drove out of the mountains back toward the fields, we looked out over the corn: straight rows, gold tassels, a curved shoreline, deep purple ocean, and the island of Ni'ihau, owned by the Robinson family, one of the large sugar baron families in Hawai'i.

In traditional Hawaiian culture, no single family owned land. Hawaiians lived in small chiefdoms before King Kamehameha conquered them all in 1810 and united Hawai'i under his rule. Around the same time, an invasion of missionaries, traders, and whalers descended upon the islands, bringing their god and greed and disease. By 1830, the population of Hawaiians had declined by more than 75 percent from an estimated pre-contact population of three hundred thousand. By century's end, American colonialists had created and controlled the sugar industry. Kamehameha's kingdom was no more, and Hawai'i became a US republic and then state.

Hawaiians, whose ancestors had lived here for a thousand years or more, long enough to have multiple names for winds and rains, were pushed from the land. A few took jobs with the sugar companies, but many wouldn't. So as plantations do, they brought in laborers from elsewhere—China, Portugal, Japan, the Philippines, Korea, and Puerto Rico. A caste system based on ethnicity was established. All leadership positions were held by White Americans. Japanese, Chinese, Filipino, and Hawaiian workers did the same jobs but were paid differently. The groups fought each other while the plantation controlled them all. By World War II it was becoming less profitable to grow sugarcane in Hawai'i. Ultimately sugarcane was replaced by corn, sugar replaced by corn syrup. The plantations sold or leased their land to agrochemical companies that continued to import labor and benefited from the plantation culture left behind.

In a way, the farms in Indiana aren't too far from plantations, though the farmers themselves aren't necessarily like plantation owners since the seed and chemical companies control when to plant, how to plant, when to spray, what to spray.

Once again I thought of how artificial fertilizers and the 'cides began as weapons of war. Companies that once produced gas chamber poisons for Nazi concentration camps carrying out genocide now provide pesticides to grow our food. I looked out over the fields of corn, the sprayer still hovering above the rows, and somewhere in the distance was the PMRF base.

"So you've got your military here," I said to Gary, "and the nameless chemical company next door."

"That's right," Gary said. "Partners."

•

At the county building in November 2013, there were the red shirts. And there were the blue shirts.

The blue shirts chanted, "G-M-O! G-M-O!" and "We are Kaua'i Ag" was written boldly on their chests. They were largely people from the west side who either worked for the agrochemical companies or had family who did. They'd been bussed in.

On the red shirts, the words "Pass the Bill" were written in bright yellow. Fern and her team had decided on the colors because they represent Hawaiian royalty.

Bill 2491 had been introduced in June at the first of several marathon county council meetings.

Sol and Fern sat in the front row of the packed room. A spillover crowd chattered outside the room.

Gary, wearing a blazer over a Hawaiian shirt, started the meeting. To present the bill to the public, he crammed six months of research into ten minutes. He talked about how difficult it was to get information from the companies. He educated the crowd on what a restricted-use pesticide was, how many of them were banned in Europe, and how, according to reports he dug up that the companies had filed with the State, twenty-two different chemicals, totaling eighteen tons, were used in Kaua'i. He pointed to a binder that contained labels for the pesticides, some of which were a hundred pages long, warning users that they kill bees, and to keep them away from children and pregnant women.

He said the bill asked for several things, including disclosure.

"For me, this is about the right to know," Gary said. "The right to know what's being used in our community in terms of chemicals, and the right to know what those impacts are. It's a statewide issue, but Kaua'i is different. We have more companies, more acres, more chemicals... Kaua'i is ground zero."

The bill also mandated buffer zones, prohibited open-air testing of experimental pesticides and GMO seeds, imposed a temporary moratorium on all experimental use of pesticides and GMOs, required county environmental impact statements, and created a process for county permitting of chemicals and GMOs.

"This bill isn't about whether GMOs are good or bad," he emphasized. "This bill is about the impacts on Kaua'i County."

"Serious, serious stuff," Gary added. "No one has studied what this means for the water, what this means for the residents around here. No one has studied what it means for the children. We don't know what's being grown. We don't know what's being sprayed. This bill attempts to find that out."

Gary talked about how difficult it was to get information from the companies. The companies told him that there were no experimental pesticides in use and yet two of them, Pioneer and Syngenta, had permits from the USDA for experimental pesticides.

County Council meetings discussing Bill 2491 packed the county building and attracted activists, protesters, industry insiders, and concerned citizens. Chants, shouts, and cheers surrounded the building late into the night. KICKA WITTE / ALAMY

After introducing Bill 2491, he received threats. He was convinced he was being followed.

"This isn't going to cost jobs," Gary continued. "Disclosing information doesn't require a company to shut down. Saying you can't spray next to a school or hospital or stream doesn't make a company close its doors and go away."

After a discussion, the council turned to the community members who wanted to speak. There were 121 names on the list.

The first was a special-ed teacher who started her testimony with the "hang loose" sign, which originated from a man who had lost three fingers working on a sugar plantation. She registered her support of the bill and then had to leave to get her granddaughter from school.

The second spoke Hawaiian. The subtitles on the recording of the meeting read: "Speaker not understood." When a council member asked her to summarize, she shifted to English. "I just passed on my testimony," she said, politely refusing.

"Okay, we'll get it translated," the councilwoman replied.

Sol was next. "What's the cost?" he asked. "What did you have to pay to spray experimental chemicals near waterways, near schools? What is the price of paradise?"

Then he paused, struggling to speak. He apologized, took two deep breaths, and tossed aside the phone from which he was reading his prepared remarks.

"I'm not going to look at that," he said, making eye contact with the council members. "We grew up together. You are my auntie. You are my cousin. This is destroying the island. It's killing the land. It's going into the water. The reefs are dying. The turtles are dying... I feel for the families on the west side that are getting sprayed by it, and I feel for the families that are scared they are going to lose their jobs.... These chemicals stay in the air, they stay in the soil, they kill the microorganisms.... [The companies] don't have a connection to the land. They don't care. They come over here for profit... and in the meantime they are dumping poisonous chemicals by the ton on the soil. But we love the island."

The red shirts applauded.

Fern gave up her spot so the legendary activist Uncle Walter could speak. He had to get back to his farms on Molokai.

"The federal government has punted this issue," Uncle Walter said. "These corporations... are using our agricultural lands, the majority of our lands to grow seeds because it is very, very profitable." He explained how when profits die off, the companies don't hang around, how on Molokai, after sugarcane and pineapple went bust, "They left us with acid soils, plastic in the ground, and you couldn't use the soils for years and years and years."

The next woman strongly opposed the bill. She was born and raised on the west side of Kaua'i and wore a DuPont-Pioneer polo shirt.

"I'm proud to say I work at DuPont-Pioneer and continue the legacy of our ancestors to support farming," she said.

Behind her Uncle Walter gave her the side-eye. Fern looked up to the heavens and sighed deeply. The woman's ancestors didn't grow corn with tractors and chemicals. They ate what they grew and they grew what they ate.

## Hawaiian Pride B 4 Pesticides

The public comments continued into a second day. The industry brought in experts from as far away as Florida. Activists came from neighboring islands and the mainland. Some of the testimonies played like a soccer match, with each fact or story eliciting cheers or boos from the spectators watching on TVs outside. Some speakers awkwardly relied on scripts, sounding like they weren't reading their own words. Others poured their hearts out and were moved to tears.

On October 13, at an eighteen-hour meeting where the glasses of water on the tables shook under the vibrations of the chant "Pass the Bill," it was time for the vote. It was three in the morning. Everyone was exhausted but wide awake.

Bill 2491 passed by a vote of six to one.

Outside on the marble steps, Fern jumped up and down to a chant of "We passed the bill! We passed the bill!" People blew conch shells, hugged, and cried.

And then, two weeks later, the mayor called Fern, Malia, and three other leaders of the movement into his office.

He was going to veto the bill. He believed that regulating pesticides was the State of Hawai'i's responsibility, not the county's, and was concerned about lawsuits from the chemical companies. One lawyer had said that Kaua'i was not ground zero for pesticide testing, as some red shirts claimed, but would become ground zero for lawsuits.

- 

The red and blue shirts once again began gathering and camping out to save their places in line before county council meetings.

On November 14 the council met to gather input and vote on overriding the mayor's veto.

Tyler Jeffries was among those who camped out the night before the meeting, which was nothing new for him. He lived on the street and often slept near the county building. A high school friend of Sol's, Tyler had had a tough life.

An agrochemical company offered to pay him and others who were living on the street $200 each to save places in line for their executives. Tyler wasn't sure what all the fuss was about, but he knew he could really use $200.

But that night, he learned about what was going on and refused to give up his spot as the crowd shuffled into the building the next morning. Instead, when it was his turn to speak, he plopped his trucker hat on the lectern, adjusted his faded blue hoodie, and shared his story about how someone tried to pay him for his spot in line.

"I know where my heart stands," Tyler said. "I watched my dad die of cancer when I was ten years old."

And then he registered his support for 2491. In a chamber where applause was not allowed, Tyler was applauded and cheered. Speaker after speaker referenced him.

Josh Mori, who had stood with Malia in Waimea, spoke as well. He recited the first law of Hawai'i decreed by King Kamehameha and written into the State law: The Law of the Splintered Paddle. The law guarantees the protection of every person, granting them the right "to lie down to sleep by the roadside without fear of harm."

"Right now," Josh read, "we can't lie in our own homes without fear of harm."

The council needed five votes to override the veto. One member who had voted in favor of 2491 had stepped down, leaving the council with only six members and a very slim margin.

Before the vote, each council member had a chance to speak. All would vote as they had in the first round—except for one. Councilman Ross Kagawa, who had supported the bill originally, would be the second and deciding vote against the override.

The room got quiet. People started crying. They were going to lose.

But then Gary made a motion to recess until the council was able to fill its missing seat. "A decision of this magnitude should be made by seven members," he said. The motion passed.

A week later, after considering nearly twenty candidates, the council voted to add Mason Chock as an interim member. Mason ran a nonprofit that promoted leadership and he often facilitated dialogue between parties with opposing views. He had even led discussions between organic farmers and the seed companies. But now he'd have to pick a side. He'd be with the red shirts or the blue. There were no purple shirts.

Mason was sworn in and given a three-hour briefing on 2491, and the vote was scheduled for the next morning.

Mason, who'd never seen himself going into politics, said it was the longest twenty-four hours of his life. Everyone contacted him with their opinions on which way he should vote. Some thought he should recuse himself because Gary's delay was corrupt. That night he didn't sleep.

"This morning I felt the overwhelming presence of my ancestors," Mason said, addressing the council and crowd. "I could feel their anguish and sorrow." He also felt the fear and pain of his sister, the mother who had stood with her two daughters and their homemade signs pleading for "Hawaiian Pride B 4 Pesticides." He would vote for his family, his ancestors, and his sister, Malia Chun.

"My hope is that my actions today are a catalyst for healing," he said, announcing his vote at the council meeting.

Bill 2491 became Ordinance 960. Malia and Fern and the thousands of other 2491 supporters had won.

Until they lost.

Just as that lawyer had promised, Kaua'i County did become ground zero for lawsuits. The agrochemical companies sued Kaua'i to enable them to continue spraying whatever, whenever, and wherever. Or as Gary later put it, they sued for "the right to spray next to schools and not to tell us about it." A federal magistrate judge ruled in their favor, saying that the State had the exclusive authority to regulate pesticides.

But the community fought back hard, and eventually—after three more years of fighting at the state legislature—won again. The agrochemical companies would be required to disclose what restricted-use pesticides were being sprayed on Kaua'i. The companies had to honor buffer zones around schools. Ultimately, Kaua'i's efforts led to a ban of atrazine in Hawai'i and a first-in-the-nation ban on chlorpyrifos.

What Gary refers to as "an epic battle between the largest chemical companies in the world" and the "little island that could" resulted in the companies reducing their corn fields by half after a few years. To Gary, the little island won.

"Kaua'i and the heart of aloha really showed in this bill process," Sol, who is now an organic farmer, said. "Even though there was a lot of division and separation, the basis of it was, you know, love. We love our land—our 'āina—and we want to make sure we take care of it."

•

### Kumano I Ke Ala Farm (Kaua'i, Hawai'i)

I wanted to learn more about 'āina and what agriculture looks like when it regenerates communities. Malia, who is a community-based youth educator and Native Hawaiian cultural practitioner, had told me if I wanted to understand what had been at stake in the 2491 battle beyond the obvious health and safety issues, I needed to visit a kalo farm.

"Planting kalo is like sticking the Hawaiian flag in the ground," she said.

She connected me with a friend, and off I went to a farm operated by Kumano I Ke Ala, a nonprofit that teaches Indigenous farming practices to at-risk youth in west Kaua'i, upstream of Waimea on the river of the same name.

"You should take off your shoes," one of my co-workers told me. "It's better to connect with the land." Everyone else in the field was barefoot. No one wants to be the guy wearing shoes in the barefoot field.

I untied my yellow shoestrings, slipped off my hiking shoes and wool socks, applied a layer of sunscreen to the tops of my feet, stood up, and looked around. I was on the edge of a square field of taro, referred to as kalo in the Hawaiian language.

The soil felt warm on the surface and then gradually cooled as my feet settled beyond the crust.

I was handed a long steel spike and then shown how and where to pierce the surface of the soil. My fellow barefoot farmers strung a line marking a row to be planted. I made a hole every foot.

The sound of steel on soil. The smell of dirt. Sweat on my brow. Each time I thrust the giant toothpick into the ground my hands slipped down its sun-warmed sides. By the end of the day, I'd have blisters on my hands and my feet.

There were around thirty kalo patches, each the size of a youth soccer field and defined by a narrow channel of water on all four sides. Thick wooden planks served as slightly bouncy bridges between them. Countless mountain streams feed the Waimea River, which flows for only twelve miles but is one of the longest rivers in the state of Hawai'i. Its water is diverted into the farm's channels. The water was stocked with fish that ate algae and provided fertilizer, and at the end of the system the water returned to the river.

We settled into a rhythm. I made the holes. Someone else set a kalo start next to each one, and others would plant them. When we were done with a section of field, it looked like we had buried hundreds of miniature rhinos, horns up and pointing in random directions.

When you are working on a physical task that isn't too mentally demanding, there's space to think. Space to converse with your co-workers. Our conversations came in fits and starts.

Each step I took in the kalo field was more intentional than any step I would've taken with shoes on. Each left a flat footprint on the surface. I took the soil with me beneath my toenails and smeared on my ankles. And I'm sure with each step I left a bit of me in the form of dead skin cells. Some folks might pay top dollar for a Hawaiian-soil foot exfoliation treatment.

When we stopped to get a drink, we walked to the head of the ditch system. The redirected river water arced out of a PVC pipe. Assured that it was safe, I knelt and drank. Rogue droplets ran down my glasses, so I tossed them aside and dunked my head into the stream.

After always being told about giardia and other untold perils of drinking from a river, there's something liberating about doing so. I got to do it several times in different parts of the world. Maybe it's because people who practice regenerative agriculture are more likely to live near clean rivers. Or maybe rivers are more likely to be clean when they have regenerative farmers as neighbors. Or perhaps people who are regenerative farmers are more likely to tell a gullible-looking writer, "Of course it's safe!"

A quarter mile downstream, no one is drinking the water. That's where the Pioneer fields overlook the river, and their bubble gum winds drifted onto the neighboring homes. Eventually the families who had sued the company were compensated for the damage the chemicals and dust had done to their homes, though not to their bodies.

At lunchtime, we gathered on the long porch off a windowless storage shed that served as a greenhouse. I ate oranges picked from a nearby tree and chatted with Malia's friend Kaina Makua, who'd come up with the idea for this kalo field and its associated cultural education program. He sat on a wooden bench with a laptop balanced on his board shorts. He'd been working on a grant.

Kalo patches like this one maintained by Kaina Makua once helped the residents of Kaua'i feed themselves. Today, experts estimate that if the barges and planes stopped delivering, the island would run out of food in a week. KELSEY TIMMERMAN

Back in 2013, Kaina had joined other traditional kalo farmers in supporting 2491. He signed a letter that stated, "We live, work, and play in the shadows of these GMO companies," and that they wanted to "ensure that local farmers, present and future, will be able to produce good, healthy food for our community, with uncontaminated lands and water." He attended the rallies at the county building, and it was during the 2491 fight that he started the nonprofit, which manages the kalo fields and works with kids. He also started the Aloha 'Aina Poi Company, which buys kalo from the nonprofit. Poi is a traditional Hawaiian food made by cooking kalo root, smashing it, and adding water until it eventually becomes a tangy, earthy paste often served as a side.

When lunch was over and everyone else returned to the field, Kaina talked to me about the farm and the valley as he reviewed a satellite image of the property; a whiteboard with sections representing each kalo patch was nearby. The length and width of each patch was written in the top-right corner of each section, and vertical Hawaiian words labeled all but one of them in black.

In that section, the word written in blue marker ran horizontally: Hope.

Hope was 115 feet long and 71 feet wide. Hope was 8,165 square feet and here, unlike the way it was in Susanna's house back in Kenya, it was right side up.

Kaina's shoulders were as broad as a paddle length. When I first met him, he asked his middle-school-aged son to explain what they were trying to accomplish through the poi company and the nonprofit.

## Hawaiian Pride B 4 Pesticides

"Tell him what *aloha 'āina* means," Kaina said to his son.

"We... want... to make this field better," the kid said, fumbling around.

"Come on, boy," Kaina said. "Why is it important?"

And then we stood in silence, letting his son find the words.

"Because it's what we had before," he told us. "And we should be able to do it again." Kaina nodded his approval.

What they had "before"—long ago—was complete self-sufficiency. They worked with the sea and the land to provide enough calories to support a population greater than the current number of inhabitants on Kaua'i.

The English translation of *aloha 'āina* is "love for the land," but Kaina said it's much more than that. It's a connection and responsibility to the land, their ancestors, and all living beings.

*Āina* isn't something that can be fully explained or read about in a book; it must be enacted and lived. His words reminded me of what the Arhuaco leaders told me in Colombia. Books and words won't change the world; only experiences can do that. Kaina's kids helped in the fields, learned Hawaiian, splashed in the river with their dog. You learn to love the land by spending time with it. It's how Kaina was raised as well.

At the University of Hawai'i he was treated differently from others because of his size, the color of his skin, the way he talked, and his rural roots. He watched other students get funding while he had to work part-time jobs and walk everywhere because he didn't have enough money for the bus. After graduating with a degree in Hawaiian Studies, he got his master's in Hawaiian language. He taught secondary school for eight years before leaving the $60,000 per year job to earn less than a third of that at the nonprofit he started.

"So do people think you're...." I stopped short and let Kaina finish my thought.

"Crazy?" he said. "Fuck, yeah. If they didn't think I was crazy, they'd all be here."

His work takes him into schools. He asks students: "When the barges stop coming with food, what are you going to do?"

Hawai'i, 2,500 miles off the west coast of the continental United States, imports 90 percent of its food. The Hawai'i Emergency Management Agency estimates that the state would run out of food in as little as one week if the barges and planes stopped. It's happened.

In 1992, after Hurricane Iniki, the barges stopped and the residents made do with hunting, fishing, and harvesting food from their gardens and trees. You'd think that would have changed things, but since then they've only grown more reliant on imports.

"When Hurricane Lane came in 2018, they stopped the barges for a couple days," Kaina told me. "People were in a panic, except for this crazy farmer and his family."

Kalo is referred to as a canoe crop, meaning that it was one of the original plants brought by the Polynesians when they came to Hawai'i. It was a staple of their diet. Pre-European contact, it was planted on thirty-five thousand acres there. In 2018, there was only a tenth as much. A teacher in urban Honolulu worked with students to restore a field. They calculated that fifteen to twenty-five hours of labor per month in kalo could yield enough calories to sustain a person. Another class calculated that Hawai'i could feed all of its people by farming 16,000 acres of kalo, which is less than 1 percent of the land ceded to the US military in 1893.

It doesn't make sense to me that Kaua'i, home to one of the wettest places on Earth, can't feed itself.

I asked Kaina to describe what this valley would've looked like before Captain Cook arrived in 1778 and before the plantations came.

As he spoke, I imagined the scene in front of me fading. Time reversing.

"Everybody in the farmlands helping each other," Kaina said. "Farming wasn't a one-person thing."

Families would go to the fields and start working at sunrise. Men and women tended the crops; children splashed in the river and irrigation channels. Conversation and laughter. And since the Hawaiians didn't have metal, the rhythm of work was kept by stone on wood, and wood and shell moving through the soil.

There were people who made cordage from plants, people who turned kalo into poi, people who made boards.

"Right around ten or eleven, you take a long break," Kaina said, "and catch your food [by hunting or fishing]. Find a place to trade.... Like for us, we would probably trade kalo for some fish.... But take at least four to five hours off. So that was your daily life, then. Everything must have looked pristine.

"That system," Kaina said, pointing downriver to the DuPont-Pioneer fields, "wasn't built for our success."

Kaina is not working to change that profit-pursuing system. He's working toward showing that another way is possible. He echoed what Dalmas had told me in Kenya about how capitalism did not fit with an Indigenous worldview.

Kaina saw what he was doing as an alternative to the plantation mindset in which people work nine to five, Monday through Friday, living for the weekends and their short vacations. Since kalo requires daily attention, he wanted to build a common building with houses around the fields for easy access. Community members would pay no rent. Housing would be free.

Even better, he said, "People ask me what my goal is in twenty years, and I say: to make sure that kalo is free. We can build something that can really change our lifestyle."

He's not interested in the efficiencies of the larger kalo farms that make use of machines and chemicals, at which one person can manage the entire operation, bringing in labor to help with planting and harvesting.

"That's not how it was before," Kaina said. "It was everybody working to make fifty acres of sustenance. Everybody eating off the fifty acres. Without the hands, that system can't work. So how do you bring that system to today's time? How do you make that happen in today's world where money is everything? We've got to be diverse and creative.

"We have to figure out how to make kalo cool again."

I asked the question I'm not sure anyone can answer: "Do people actually want to farm?"

## Hawaiian Pride B 4 Pesticides

"As human beings, we shouldn't be confined to a four-walled building," Kaina said. "That's not natural. We go crazy. But when you put us here, and we put our hands down in the soil instead of up and asking for stuff, you put your mana [spiritual energy] in your 'āina, which becomes food. That's why we do what we do. Our efforts are not only to build sustenance, but health. Health is laughing. Health is sweating. Health is constantly moving and trying to figure stuff out.... Health is feeling the vibrations of 'āina on your feet."

•

I was trying to explain the topic of this book to someone I met while traveling. I mentioned several Indigenous groups I'd spent time with, including the Arhuaco, Maasai, and Tupinambá, and how important I found their work.

"So, you are buying into the noble savage thing?" he asked.

I bristled. The racist concept of the noble savage views Indigenous people as untouched by modern life and therefore as romanticized symbols of natural human goodness and purity. Nothing like stereotyping a diverse, global community and lumping them all together. The idea of the noble savage also freezes a people in time. As if being Indigenous means to live in the past and not adapt, react, and imagine an ever-evolving future. How can there be a future for Indigenous people if we only think about them in the past?

Kaina isn't trying to lead his community back to the 1800s. Instead he's acknowledging that his Hawaiian ancestors over thousands of years had settled onto a way of life that can inform and enrich his life, the life of his children, and future generations.

One way he promotes his work is through Instagram videos. Every Wednesday, Kaina and his brother make a Hump Day video. As they clean up at the poi company, they pause to perform choreographed dances to songs like Rick Astley's "Never Gonna Give You Up," Prince's "Kiss," and during the holidays, Mariah Carey's "All I Want for Christmas." They wear wigs, glasses, and costumes to fit each song.

I let Kaina get back to his grant. He told me that before I left, I should go to the end of the field, to one of the most important parts of the farm: the kalo patch that's the farthest downstream. The plants here were near maturity and their leaves had formed long, variegated hearts. I wasn't sure what

Employees and volunteers of Kumano I Ke Ala plant a patch of kalo. The organization reconnects residents and youth of West Kaua'i with traditional farming practices. KELSEY TIMMERMAN

## Hawaiian Pride B 4 Pesticides

I was looking for. And then I heard it: the sound of the water leaving the last ditch, heading back to the Waimea River. What nature gives is returned.

And those gifts come with responsibilities. Kaina and traditional farmers like him, the original regenerative farmers, honor the legacy of what has come before—the upstream—while accepting responsibility and giving back to what comes after them—the downstream. Sometimes that looks like bending to plant a kalo patch, and other times that looks like standing up in a courthouse and speaking out.

Industrial agriculture is about extracting, taking. Regenerative agriculture is about giving.

erate

# Part III: Regenerate

# Chapter 7
# River of Life

# Bluffton, Georgia

**Regenerative Agriculture Restarts the Cycles of Nature**

Will Harris had barely set foot in a grocery store his entire life, but there he was in his cowboy hat cooking and slinging meatball samples in Whole Foods near Atlanta. His boots on polished concrete floors. These customers were his customers—or at least he hoped he could educate people about the benefits of grass-fed, rotationally grazed beef and win them over, one meatball at a time—but they weren't his people. Growing up in rural Bluffton, Georgia, he had never met anyone who'd fit in with the Whole Foods crowd.

Will had rarely been to a big city until he was drafted into the military. He felt out of place there, too. At one point he was told to take off his clothes and line up for inspection. When they were asked to drop their robes, he noticed everyone else was still in underwear, but he was buckass naked. He was drafted late enough that he never had to go to Vietnam, but he did do several tours at Whole Foods. By 2008, he personally sold the first grass-fed beef they'd ever carried, visiting all sixty stores that stocked his products. In the beginning he sold them every pound of beef he could produce.

Will is the real deal. Two hundred thirty-plus pounds of fourth-generation Georgia cowboy. Nothing is for effect. His accent—not

PREVIOUS SPREAD: Regeneration asks us to work with the cycles of nature. WHITE OAK PASTURES

OPPOSITE: At White Oak Pastures cattle, sheep, and chickens all play a crucial role in growing healthy soil. WHITE OAK PASTURES

quite as twangy as his father's, who sounded like the cartoon rooster Foghorn Leghorn—seemed a little too thick at first. But I quickly realized the way he speaks is as genuine as his meatballs are delicious.

The hard part for Will wasn't standing over sizzling samples in Whole Foods. The hard part was explaining to these people how he produced the meat at White Oak Pastures, his five-thousand-acre farm, and why it was better for them and the planet.

"I really appreciate what you do," a customer said. "I think it's great, but I gotta say that I really don't understand how you can have a baby calf born on your farm, and care for it for two years, and then kill it and eat it."

That didn't anger Will. But it puzzled him.

He wanted to explain, "Killing him and eating him doesn't bother me a bit, ma'am."

But that sounded flippant, and Will isn't flippant. He recognized that there was a disconnect between his life experiences and the customer's. After getting that question again and again, he figured it out. The customers' relationships with animals were framed around having a puppy or some other cute, furry pet.

So he learned to meet them where they were, and say something like this: "Ma'am, I've always had dogs. They're a companion animal. I feel the same way about my dog as you feel about yours. But I also have working animals, guardian dogs that look after the sheep. They ain't got names. Still, I care for them and I love them."

If the shoppers were still with him, he'd finally get to the point, saying, "Then there's my livestock. My livestock is a river, not a lake. I don't love the individual; I love the flow. I love seeing calves being born so I can raise them. And I love seeing happy, healthy cows go off to the slaughterhouse to make way for new calves. I love them all."

Sometimes he'd win them over and other times not. Sometimes people in grocery stores don't want a big dose of truth and authenticity. They don't want to be reminded that something died so they could eat it.

The question that did boil his blood was this: "Oh, I've read all about y'all, and I just love what you do. But what can you do to make it affordable to poor people?"

"Well," Will would say, "you've taken two of the most complex problems in the world: the broken food system and income inequality. And then you put them together and told me to fix it. I'll tell you what I'm going to do. I'll fix the broken food system and you fix income inequality. How about that?"

I doubt his suburban Atlanta customers knew that White Oak Pastures is located in one of the poorest counties in all of the United States. And what Will didn't know in 2008 was that his farm would bring new life to Bluffton and offer hope to other eroding rural communities that might be watching.

•

I rode shotgun next to Will in his Jeep Wrangler, my knee bouncing off an upright AR-15 held fast by a bracket on the dash. Spent casings jingled at my feet as we bumped through a pasture.

"You got some shells on the floor here," I said. "You firing out the window?"

"It could happen," Will said.

"What are you shooting at?" I asked.

"Anything that doesn't have the best interest of White Oak Pastures in their hearts," Will said.

There are a lot of hearts on Will's farm—one hundred thousand on any given day, by his estimate. There are cow hearts, pig hearts, sheep hearts, hearts belonging to goats, rabbits, chickens, guineas, turkeys, and ducks. And that's just the hearts involved in his fifty-six enterprises—from leatherworking to eggs to bees to online sales to crafts to a restaurant to tours to

Will Harris is a fourth-generation farmer and owner of White Oak Pastures. His father had taken the farm from an organism to a factory, and now Will is taking it from a factory to an organism.

WHITE OAK PASTURES

## River of Life

a slaughterhouse—operating on the farm. That doesn't count the other human, animal, insect, and plant life that White Oak Pastures attracts.

I got out and swung a gate open. It had rained the previous night and Will gave the Jeep a little extra gas to run up out of a deep puddle. Some people, he said, drive a Jeep to feel strong. He drives one because he doesn't want to walk home. His cowboy hat keeps the sun off his face and his mostly bald head—except for one tuft of white hair front and center. His gun brings protection and mercy to his livestock. His suspenders hold his pants up. And sometimes he has to pronounce words precisely to cut through his accent for me to understand what he's saying.

I closed the gate carefully, because I'd seen signs on other gates that read: "Keep the Damn Gate Closed—Scott Cleveland." I wasn't sure who Scott Cleveland was, but I didn't want to be on his bad side.

Will's daddy used to say, "The best thing you can put on your land is your feet, and the best thing you can put on your animals is your eyes." That's what Will was doing. If he saw something that needed to be addressed, he'd stop the Jeep and shoot off a text to one of his twenty-six managers.

He angled toward a herd on the edge of a small forest, which I recognized. The day I'd arrived and settled into my cabin, I had gone for a walk. After visiting this same herd, I'd decided to bushwhack my way back to my cabin. The cows had watched as I disappeared into the trees, which quickly became a swamp laced with vines and thorns. Orb-weaver spiders had constructed webs so thick I could hear them snap when I walked through them. As nighttime and thunderstorms neared, eventually I gave up on jumping across streams and puddles or worrying about snakes or alligators and just crashed through the vegetation.

"I can't believe you walked through Devil's Branch," Will said. Of course, if I had known it was called Devil's Branch I probably wouldn't have. "It's amazing your arms aren't all cut up. I mean, I've done it a bunch of times, but usually to get something. You needed a machete."

"It was like being in the rainforest," I said.

"Well, this is the semi-tropics," Will said. "This land longs to return to the jungle. When you clean it up, that's what'll happen, unless you do one of five things: You can spray it, burn it, till it, mow it, or graze it."

Of course, Will favors the latter because as he has put it, "If [soil] microbes could be heard screaming, we would never practice these first four."

Will turned off the Jeep and watched the cows graze. Cattle egrets flew above, walked around, and sometimes sat on the backs of the cows. The birds eat the grasshoppers the cows scare up with each step and also the flies attracted to the manure.

"It's like an African safari," I said.

"A lot of people say that," Will said. "People don't understand the importance of going to your pastures. My guys think I'm magic because I can ride by the pastures and quickly spot five things that are wrong. I'm not magic."

But he does have a great eye, trained by decades of observation.

## River of Life

He explained, "The cows when they are thirsty move and mill a certain way. When I see a cow by itself, or one walking funny, I know something is off. When I see that a cow reached through the fence and bit the grass—I can see the fence ain't on. I can do that at twenty-five miles per hour because I sit here at zero miles per hour and understand what normal looks like."

White Oak Pastures is one of the leaders of the regenerative agriculture movement. Will and White Oak Pastures have been featured in documentaries and highlighted by major news outlets from CNN to the *New York Times*. He's trained hundreds of young farmers on his land, served as President of the Board of Georgia Organics and as Beef Director of the American Grassfed Association, and has received recognition from around the world for his humane animal husbandry and regenerative practices. Perhaps the only other comparable American farmer—who works regeneratively on a similar scale and is as well known—is Gabe Brown, a rancher in South Dakota. Both of them originally learned to farm with chemicals.

"Gabe's my best friend," Will continued. "Gabe wisely said that when he was an industrial farmer, every day he walked outside looking for something to kill, and now he walks outside, looking how to make things live. That's very profound. I didn't come up with it, but I plagiarize the hell out of it. I love the term 'teeming with life.'"

I had seen this from farm to farm on this journey and in my own small-farm adventures. Hate ticks? Fight them with guineas like Mark Shepard. Hate moles digging in your yard like me? Don't kill the snakes. "When we do what we do, it teems with life," Will continued. "People piss me off when they have this Walt Disney

The formidable Scott Cleveland, one of Will's top cowboys, watches the cattle clop across the road in anticipation of greener pastures. White Oak Pastures has become Bluffton's largest employer and is credited with helping to revitalize the rural town. WHITE OAK PASTURES

## River of Life

version of what teeming with life means. They are thinking deer and bunny rabbits. But we have seven poisonous snakes here. Teeming with life means insects, reptiles, rodents, bushes with thorns. Teeming with life is the way it's supposed to be."

We joined a dirt road, making our way to the northernmost edge of the farm and toward one of the sexiest things on the ranch—a cattle crossing. Will joked that the event always drew a crowd of employees, more than actually were needed to help the cattle make their way over the highway from one paddock to another.

His eyes always searching, Will spotted a bald eagle on the top of a dead tree. "That's a juvenile," he said.

The farm has one of the highest populations of bald eagles, more than eighty, in the state of Georgia. The eagles have eaten hundreds of thousands of dollars' worth of chickens. And since the eagles are federally protected, there's not much to do other than to learn to live with them. The chickens run free in moveable coops. The poultry team began to install temporary fencing in the pasture around the coop, which has reduced the killings by disrupting the raptors' flight path. Personally, Will loves to see the eagles—a sign of a healthy ecosystem—although he curses the government, which has yet to reimburse him for his losses at the talons of the protected predator.

We passed the organic garden plopped in the middle of a pasture, which grows veggies of all sorts for the White Oak Pastures restaurant. Then there were the beehives and the start of Will's food forest of sapling peach, pear, apple, and plum trees. We came across a team of people moving fencing for the sheep on the edge of the road. I live on twenty acres, Will lives on five thousand, and it turns out we have the same number of lawn mowers: one. But he does have two thousand sheep. When you have sheep, you don't need mowers—and you certainly don't need Roundup.

He pulled over alongside one of his neighbor's fields of corn, only a mile but worlds away from the organic garden and food forest. He turned off his Jeep and but for the cicadas welcoming the rising temperature of the day, it was quiet. There were no birds singing, no cows mooing, no sheep or goats bleating, fewer bugs flying through the air and adding to the chorus. It was not teeming with life.

People have come from around the world to learn from White Oak Pastures, an oasis completely surrounded by monocrops of industrial agriculture.

Since 1995, Will has dedicated himself to improving the land, and I expected him to have strong opinions about his neighbors' adherence to industrial, chemically intensive agriculture. Instead, he spoke admiringly of their perseverance and business sense. "The thing you have to understand is that the farmers who are still farming like this are the absolute best at what they do," he said.

I found relief in his words. It was possible to look down on the industrial agriculture system and not look down upon the people—like my Midwestern friends, family, and neighbors—who were stuck in it. In fact, Will understood why they continued to farm that way.

"Most of these guys are locked into the system. They've got $750,000 harvesters. I know folks who have three of them. Most of them are partners in grain elevators or peanut warehouses or peanut shellers. Most of them do okay financially."

## River of Life

The industrial system is efficient at one thing: growing a single crop. Conventional farmers have a chemical solution for most unexpected problems, irrigation for droughts, and federally subsidized crop insurance to make up their losses. Industrial agriculture exposes farmers to fewer financial risks than Will's method. But while his system isn't as efficient, it is more ecologically resilient. He doesn't have crop insurance, but a diverse system of fifty-six different enterprises provides a safety net. He doesn't have chemicals, but animals and heaps of compost fertilize his land, pigs clear the woods, and sheep and goats clear his fence lines. He doesn't have irrigation, but the living roots in the ground at all times allow his land to retain more water.

Living roots prevent erosion by physically holding the soil, but their impact is much greater than that. They support an entire underground ecosystem by pumping carbon down to where it feeds a web of life that holds even more soil. Will's deep, living roots put sunshine and cow patties to work to build more soil.

There's a place on the farm where one of his ditches, which drains two thousand acres, meets one of his neighbor's ditches, which is fed by one-tenth the area. During large rain events, the neighbor's ditch is so full of red topsoil it looks like strawberry milk, and Will's looks like a normal tea-colored stream.

Will used to farm just like his neighbors. His daddy had bought into industrial agriculture in the '60s. Will was a boy when his father was given a fertilizer sample. It made the grass grow faster. What wasn't to like? His daddy used to say that you could never have too much money or grass. Will went to the University of Georgia and learned how to further perfect the industrial system, which he was a part of for twenty years.

"Daddy was in charge of the first transition—from an organism to a factory. And I'm in charge of the second—taking it from a factory back to an organism."

Will doesn't place any blame on his father for farming industrially, and he doesn't claim moral superiority, saying, "I didn't figure out how to do regenerative agriculture. I figured out that industrial agriculture wasn't right. Not because I was sensitive, smart, or insightful, but because I was an abuser. If the label said give one cc, I'd give two. If it said use a pint, I'd use a quart. If you were supposed to put this many in a pen, I put in more. I'm not proud of that. I tell people that if you drank a shot of bourbon

After more than three decades, Will is one of the elder statesmen of regenerative agriculture. White Oak Pastures launched the nonprofit Center for Agricultural Resilience to educate and inspire other farmers around the country and the world.

every night, the ills of alcohol won't be evident, but if you drank a quart of bourbon every night.... I was like that. An abuser. The unintended consequences were noticeable."

Will didn't share any further specifics about his environmental abuses, but he didn't have to. Most of us who live in rural America have accepted the land's lifelessness as normal.

Will wanted to buy the field we stared at. He always wants to buy more land. Over the years, he's expanded White Oak Pastures from the family's original eleven hundred acres to nearly five thousand, which is by far the largest farm I visited on this journey, but not out of the ordinary for farms in the US. When he's in his office, he constantly has Google Earth up, looking at the properties next to his. His daughters call it his "land porn." But he only buys land he can fix. If he bought this field, which he calls "a dead mineral medium that has been farmed to death," he'd winter the cows on it and give them a lot of hay, something he calls "hay bombing." The cows would spend six weeks peeing, pooping, stepping on the hay, which jump-starts the microbial life in the soil. The next summer, he'd use a no-till drill (no plowing involved) to plant perennial grasses or maybe annuals for the first year. Once the pasture was established, he'd begin rotating cows through the following year.

And even with this tried-and-true method, in some pastures he's had in his system for four years you can still see lines where herbicide was last sprayed.

"We don't know what we are doing to our land," Will said.

It was time to move on and get to the cattle crossing. We left the field and headed up the road. After a while, we found muddy pickup trucks blocking the road and others pulled over in the ditch.

The herd of nearly two thousand cattle were bunched at a gate, impatiently mooing the fresh grass moo to five cowboys. I wished that Leshinka and Dalmas were here to see so many cows in one place. I think they'd have a lot to talk about with Will.

"If you come stand right here," Will said, "you won't get run over."

Only one of the cowboys wore an actual cowboy hat. He was a younger, slimmer version of Will and ran over to greet us. I could tell before Will introduced me that he was Scott Cleveland, cattle production manager, former Marine, and a man who seriously wanted the damn gates to be closed. Will had told me Scott's entire backstory. It was a colorful one, told in confidence, which culminated with Will walking in when Scott was interviewing for a position in the slaughterhouse. Since cowboy recognizes cowboy, Will snatched Scott aside and asked if he'd like to work with the herd instead.

"I don't know a goddamn thing about cows, Mr. Will," Scott had said.

To which Will had responded, "That's exactly the kind of cowboy I'm looking for." Will liked to train his own.

Will's hiring and management style is every bit as boots on the ground as his farming. White Oak Pastures had 180 employees and it seemed as if he knew all their names, even the folks who'd just started. More than that, he knew all about them. There was a New York City musician who had toured around the world, a man from Haiti interested in holistic management who was the second-best fence guy on the farm (modesty precluded

## River of Life

Will from saying who the first was), and a homeopathic practitioner trying to gain a better understanding of nature. For some employees, their jobs represented a second chance after they had fallen on hard times or the wrong side of the law. Some of the stories involved White Oak Pastures helping out beyond any financial and social measure I would have expected. I contrasted Will's connection with his employees to those at the Tyson meatpacking plant where workers, mostly immigrants, had been ordered by the President of the United States to go back to work during the first wave of the COVID pandemic, and how their bosses had placed bets on how many of them would die.

Will told me that some of his management team had tried to convince him that White Oak Pastures should create an employee handbook. Will wasn't having it. A handbook would be full of policies that sought to treat everyone the same, and Will thought that meant they wouldn't be treated fairly. Treating people fairly meant taking into account each person's individual life circumstances, experiences, strengths, and weaknesses. Will didn't manage his pastures or his animals as interchangeable objects, so why would he manage his employees that way?

Scott Cleveland had been with Will for three years by now, moving up the ranks. "Nice to meet you," I told him, shaking his hand, and then he and Will talked about the grass and the rain. A congregation of cattle egrets flew above the cows, waiting to descend upon the fresh pasture as well.

Finally, Scott ran over to the last gate. "Let's roll with it!" he hollered. He opened it and ran across the road. The cows followed him, their hooves clopping and slowly spreading mud on the pavement until the yellow center stripe disappeared.

Will watched, assessing his river of livestock, and maybe with more than just a touch of pride. "See how shiny they are?" he asked, his thumbs looped in his suspenders. The way the sun shimmered on their hair reminded me of the cattle at a county fair that the FFA (Future Farmers of America) kids trim, groom, and bathe before walking them around the ring before the judges.

"What breed are they?" I asked.

"They've been mongrelized to hell and back," Will said, without taking his eyes off the herd. "They go back to the cattle my great-granddaddy brought here in 1866."

After three minutes, the cows dispersed in the fresh pasture, the cowboys shut the gates, and the egrets descended, some coming in for a direct landing on the cows' backs.

"My cattle have never been on a trailer," he said. "They are born here and walk anywhere they've ever been, and then they die here."

That wasn't always the case. When Will's daddy ran this farm using the industrial model, the cows would graze for a year before being shipped off to the feedlot to be fattened up on corn before being slaughtered. Then one day in 1995 when Will was loading cows on a trailer for that trip, it all crystallized for him. He'd spent time with each animal, observing it, making sure it was healthy, that it had all the nutrients and water it needed, plenty of grass and corn to eat, plenty of antibiotics and hormone implants. Now the animals were scrunched into a trailer, panicking and pissing and pooping on each other. And for some reason that day, it struck him. "This shit sucks," he thought. That

## River of Life

was the beginning. It started with animal welfare, and then he began questioning the whole approach to farming.

When Will explained the system he inherited and then perpetuated for two decades, he enunciated three words like a spoken-word poet, each one dripping with venom: "Centralized. Industrialized. Commoditized."

Centralized. Very few companies controlled the agrochemicals he relied on for his pastures. Very few companies controlled the processing of his beef and the price he'd get for it. Corporations that faced little competition put the squeeze on farmers. The cost of chemicals entering their farms goes up and the value of the cattle leaving them goes down.

Industrialized. The goal was to produce food as cheaply as possible for as much profit as possible. So quantity and efficiency reigned supreme, favoring large operations able to multiply small profit margins across an unknowable amount of acres and animals.

Commoditized. No matter how much care and pride he took in his cattle, how much he loved the flow of the river, it ended in an ocean of beef. It didn't really matter how the cows were raised or even what the beef tasted like because it was dumped in with a bunch of other beef. And the price of that beef relied on the price of another commodity: corn.

And it's all gotten worse. As of 2021 Tyson, Cargill, and two Brazilian-owned companies, JBS and National Beef Packing Company, dominate the beef industry, with 85 percent of the American market share. Americans are eating more beef than ever, and during COVID, the price of beef skyrocketed 40 percent

Learning from his own observations and Allan Savory's teaching on holistic management, Will mimics nature by bunching and herding his cattle. WHITE OAK PASTURES

while certain cuts went up 70 percent. So, the four meatpacking companies are making record profits—but those profits aren't making it to ranchers. In 2020, ranchers earned 37 cents for every dollar they spent, leading a Missouri farmer to tell the *New York Times*, "You put your heart and soul into something, and then you lose your ass."

So maybe Will got out of the industrial system at the right time. First he stopped using hormone implants, a regular diet of antibiotics, and corn feed. In 2003, he quit chemical fertilizers. For a few years his pastures did not perform well and he lost money since he had to supplement with hay. Then he started to add sheep and goats and chickens and turkeys. And then he really took a big leap. To complete the circle, he built his own $7.5 million, USDA-inspected slaughterhouse.

•

"That was a million and a half pounds of beef that walked across the road," Will said as we got back in the Jeep and headed toward his office in Bluffton.

"How much will it sell for?" I asked.

A surprisingly complicated calculation ensued, taking into account the cost of feed and land (some of which needed to be restored to the tune of $500 per acre after Will bought it from a neighbor). Will landed on an estimate of $2 million. Raising grass-fed beef has lower chemical and medical costs, but it also takes longer than the industrial-feedlot model. Will's cattle weigh about a thousand pounds when they are slaughtered at two years of age. Industrial, grain-finished cattle reach weights of over sixteen hundred pounds and are slaughtered eight months earlier.

## River of Life

"There's a reason there aren't a lot of rich farmers," Will said.

White Oak Pastures was one of the first farms to be certified humane by the Rodale Institute. Will thought the certification was the answer to his problems, and that with it he would no longer have to educate consumers on why his holistically managed grass-fed beef was different. He could focus on his land and not have to stand in Whole Foods cooking meatballs.

But he was disappointed. In 2017, the food industry erupted with more than one hundred labels and certifications, which were often confusing, if not outright bullshit. For instance, beef labeled "pasture raised" may not have ever seen a pasture. Any meat from anywhere in the world got the "Product of U.S.A." label if it underwent any processing in the country. Until 2017, American farmers like Will held 60 percent of the "grass-fed" market, but their share dropped to 20 percent as that label—along with "natural," "grain-finished," and "humane"—lost their meaning.

Instead of providing clarity, Will felt that labels and certifications brought confusion. Even though he sees validity in some of them, he feels like they contribute to the greenwashing—making claims that help consumers feel virtuous about their choices, even when they're not based in reality—one of the regenerative movement's biggest threats.

On a 2022 appearance on Joe Rogan's podcast, Will accused Whole Foods of greenwashing and undercutting his products with meat labeled as ethical and humane, grass-fed, pasture-raised American beef, but was often none of these things.

In 2023, Will's long partnership with Whole Foods came to an end. In a letter to Whole Foods' customers announcing that products from White Oak Pastures would no longer be available, Will wrote that "[t]he same Whole Foods Market that once embraced small, local agriculture is now focused on the same efficiencies that keep farmers from transitioning to regenerative systems."

Will looked at the loss of one of his largest customers (and early on, his only customer) as another tough lick in a long line of tough licks he's had to overcome. He admits that the transition from industrial to regenerative, from factory to organism, wasn't an easy one. His daddy had run the farm debt-free, but Will had to take on debt to buy land (grass-fed cattle require more space), build the slaughterhouse, and expand his operation. There were other stressful times where he'd work all day, stay up late, drink a bottle of wine, fall asleep for a few hours, and then wake up to do it all over again. His wife, a schoolteacher, had faith that he had things under control. He never told her, but there were times where he could see the path he was on and it looked like they could end up in a rented mobile home in Bluffton. Even with five thousand acres, fifty-six enterprises, and being seen as one of the leaders of regenerative ranching, it's still a possibility.

Economist Raj Patel and National Family Farm Coalition president Jim Goodman have written that "those who want to farm with dignity in the web of life plead a case for which there is no business logic." Will seeks to prove sentiments like that wrong, however he fears they may be right.

The White Oak Pastures management team sees a future where everything will be paid for, as long as they can keep Will from giving into his land porn desires and buying all of Clay County.

**River of Life**

Will doesn't think the farm will break even during his life, but he's not quitting anytime soon.

We visited a property that Will acquired last year. We got out and walked down a slope toward a wooded area.

"Is that bandanna there for a reason?" I asked, pointing to one tied to a limb of a magnolia tree.

"Yeah, I was telling my daughter where to bring my grandson to look at the beaver," Will said.

Will and I had discovered that we were both "beaver believers." I introduced Will to the term, which he adapted to Brotherhood of the Beaver. Anyhow, we are both amazed by the work beavers do.

"There must be two to three hundred acres backed up here," Will said. "There's not just one dam. There's a series."

We walked to the water's edge. The place was so teeming with life it hummed, croaked, splashed, and clicked.

"I love this place. I'm probably going to put me a little cabin right there," Will said as he pointed up the slope toward the handkerchief.

"To most farmers, this place is worthless," I said.

"Financially it is," Will said, "from an income-producing perspective. And you have to pay taxes on it. But I do love this place. I'll tell ya, I'll never retire."

We followed the stream up to the headwaters, a natural spring at the base of the seventy-eight-foot bluff that gave the town of Bluffton its name. Will's grandmother's brother built a concrete swimming pool at the bottom of the cliff in the 1920s. There had also been a dance pavilion down there where Will's granddaddy courted his grandmama and his daddy courted his mama.

We looked out over the lush canopy of a subtropical forest. I wouldn't have been surprised if a monkey had swung by on vines or a jaguar lurked out through the brushy understory.

"You know," Will said, "there's a question that you haven't asked me that you should ask."

"What's that?" I asked.

"How to define regenerative agriculture," he said. "I've got the definition."

There was a reason I hadn't brought it up. By this point, I kind of knew what to expect. The farmers I talked to in the United States often used definitions focused solely on soil health and carbon sequestration. Worthy goals for sure, but they never quite captured what I had experienced in Kaua'i, Kenya, Brazil, and Colombia. And since Will was featured in a documentary series titled "Carbon Cowboys," I expected him to use similar language.

Plus, he was one of the few people who hadn't asked me for my own definition of the term before he would talk with me. I'd hoped he wouldn't get around to it, since I was still working on it. I had to admit that the more I experienced regenerative agriculture, the less I thought it could be defined.

"I don't get why people are struggling so hard," Will said. "It's very easy and very intuitive. The cycles of nature create the abundance. All that oil and coal in the ground is the abundance of the dinosaur age. During that era, all the cycles of nature were working optimally and all that energy was trapped below the ground. And fast-forward to when Europeans got here.... All the cycles of nature were operating optimally and supported a population of Native Americans."

I could see where he was going. He continued, "And then Europeans got here and started, in a very puny way, breaking the cycles of nature with wooden plows and oxen. And then World War II happened and all this new technology was invented. All this passion and focus went into making tanks. The first pesticides came from tear gases.... The munitions effort made cheap ammonia fertilizer."

"Implements of death," I added. "Meant to kill people."

"Literally," Will said. "And then those young men came back from the European theater where they'd been driving tanks. They didn't want to plow with a mule. They wanted a tractor and to turn the land deep. And so what that technology did is, for the first time ever, humans could break the cycles of nature."

I imagined standing on the edge of the bluff after the war and hearing the echoes of kids swimming in the pool being drowned out by a tractor in a nearby field for the first time.

"The cycles of nature... first of all, I don't think we're smart enough to even know them all. There's a water cycle. We know there's an energy cycle. We know there's a carbon cycle. We know there's a mineral cycle. We know there's a microbial cycle. And there are others I could come up with, like community dynamics. But we broke them. Technology broke them."

"Every one of them?" I asked.

## River of Life

"Every one we recognize," Will said. "Regenerative agriculture is simply restarting the cycles of nature. You know, industrial, reductive, high-input agriculture broke the cycles of nature, and regenerative agriculture is making them work again. And some of us have figured out how to do it. And those of us who have figured it out probably aren't the smartest ones. We're just humble enough to look."

Regenerative agriculture is restarting the cycles of nature. I liked it. It acknowledged that the cycles had been stopped, and the fact there were cycles—plural—acknowledged that nature was complex. It offered hope that humanity could play a role in getting things going, and it sounded oddly familiar. It took me a bit to realize it, but what Will said harmonized with what Freddy, the Arhuaco mamo, half a world away, had said: "Nature... is perfect. We are the imperfect ones. We have broken the rules of the Earth, and its perfect cycle." Will's view also reminded me of Carmen Fernholz standing in his field of Kernza, Mark Shepard in his food forest, and so many of the others I met. It's as if no matter where you are, if you observe the natural world around you long enough, you are humbled into working with nature and not against it.

"So often people want to make regenerative agriculture a checklist," I said.

"Yeah, people want an instructional manual, and there ain't one," Will said. "I know how to do it here. I'm attuned to this."

Many of the agrochemical 'cides have their origins in war. Implements of death are now used to grow our food and fight against nature. W. G. THAYER / LIBRARY OF CONGRESS

"Is it a spiritual thing for you?" I asked since what he was saying reminded me of what the Arhuaco had told me about books not being the answer, about instead sitting, listening, and observing a place.

"Yeah," Will responded. His voice was almost a whisper.

"In what sense?" My question seemed jarring in comparison.

"Shhhh..." Will said, putting his finger to his lip. "We in church."

*Hum. Buzz. Rustle. Tweet.*

"I think it was all paradise," he continued. "Whether it was tundra, or rainforest. That whole Garden of Eden thing, I think that's what it was all about. The apple was technology and we ate the shit out of it."

*Howl. Flutter. Shake. Chirp.*

"Hurricane Michael kicked our ass. We're still... we'll never be the same." Will's voice cracked with emotion. "Ah, yeah, we will.

"I think it's real pretty," Will said, looking over the bluff. "Well, it was really pretty until Hurricane Michael."

The Category Four storm went through in 2018 and wreaked havoc. No people on the farm were hurt or worse, but the animals weren't so lucky. Some were killed and others were injured. The hurricane did $3 million in damage to White Oak Pastures.

"For me, the greatest personal loss was the woods. That cost me nothing. But these woods will never look the same in my lifetime. Hurricane Michael kicked our ass."

**River of Life**

"When you talk about it," I said, "I feel like you get a little emotional. Is it the loss of the woods that gets you the most?"

"Yeah," he bellowed, and once again got quiet, but then smiled. "You are mistaken... cowboys don't get emotional."

•

Bluffton, Georgia, with a population of barely a hundred people, has the feel of a recently unabandoned town. The houses looked like they'd been newly repaired and are mostly cared for but not yet quite lived in. For most of Will's life, Bluffton, Georgia, had been dying. The general store sat empty for more than fifty years, a time when the only thing you could buy in town was a stamp at the post office.

Then in 2016, White Oak Pastures bought the building and fixed it up to sell their products, a few grocery items, and serve food. Now there's also an RV park and other lodging. People are moving to Bluffton. In 2023, the Georgia Department of Economic Development named White Oak Pastures a "small business ROCK STAR" for revitalizing Bluffton. Turns out Will's way of fixing the broken food system actually helps fix the broken economic system, too.

Will's office was in the small old Bluffton courthouse. Snakeskins are stretched out on the white pillars of the porch. They are tacked on the inside, as a reminder of the time Will's daddy was bitten by a snake and had to spend a few days in the hospital. A Lakota man once told Will that snakes have a collective memory and that if you kill one, they all remember.

Will told him, "I believe that, because my daddy got bit by a snake and my grandaddy got bit by a snake, and I ain't forgot.

I don't kill non-poisonous snakes; I protect them. But I kill the poisonous ones. I kill the shit out of them because the Harrises got a collective memory, too."

Inside, Will sits at the judge's bench, facing the entrance. Bullwhips hang on meat hooks on either side of the door, and sketches of each generation of Harris who worked on this land, starting with Will's great-grandad, hang above. A visitor once surveyed the sketches and observed that the Harris men got leaner and meaner with each generation. But things changed after Will's sketch. Instead of one person, there were two. And instead of serious-looking men, there were two smiling women—Will's daughters.

Will's daddy ran the ranch with only three employees. One of the unintended consequences of Will's shift away from the industrial model was that it created plenty for his kids to do on the farm. But Will had a rule that they had to spend a year working off-farm before they could return, if they wanted to. Two of his three daughters moved back to Bluffton to work on the farm and to raise the sixth generation.

Jenni is the Director of Marketing. As a gay woman, she never imagined living in Bluffton after going to college. She thought she had to make a decision between being personally happy or professionally happy. But her parents were nothing but accepting, and she and her wife, Amber, were welcomed into the community. She's proud that Bluffton has become a bit of a rhinestone in the Bible Belt and jokes about hosting a pride parade. Jenni and Amber are proud mommas to their son, Jack.

Jodi works as the Director of Farm Experience, overseeing the on-farm lodging and the general store. She grew up rodeoing

and never imagined living anywhere else. Her husband, John, is from the suburbs, but Will recognized the cowboy in him and John is now the Director of Livestock.

The management team is a family affair, with a non-family scientist and a non-family accountant thrown in the mix. Once a week they gather at the courthouse for a meeting.

I joined them as they put together the details to launch the Center for Agricultural Resilience, a nonprofit that would bring in farmers for four-day courses on rethinking the systems of agriculture and restarting the cycles of nature on their own farms. The goal is to use lessons learned at White Oak Pastures to guide

Generations of Harris farmers—all unsmiling men until Will's smiling daughters joined them—look down from above the door of Will's office in Bluffton's old courthouse. KELSEY TIMMERMAN

and inspire farmers to grow their own regenerative farms in a way that rebuilds their land and their communities.

Having more people connected to the land and therefore to each other fights against the threat my Ohio Farmer friends see that someday land will all be owned by large corporations and billionaires. Which, in fact, is happening: In 2022, Bill Gates—yes, him again—became the largest private owner of farmland in the United States, owning more than 240,000 acres.

How could anyone know or have a relationship with so much land?

Of course, Will had an opinion. "Land is precious," he said. "It may be more precious than anything. I hate to see someone who has no idea what to do with it be put in a position to control it.... Gates believes that the cure for all problems is more technology.... I have a deep understanding of how misapplied technology is responsible for most of the land-management problems that we're experiencing today."

Will knew every acre of his farm and he was passing that knowledge onto his children and employees. Just a few years ago, the farm couldn't run without him. He couldn't have afforded the time it took to show someone like me around. He was irreplaceable—which was unnatural. In nature, nothing is irreplaceable. The team has all worked together to get to this point—weekly meetings where every voice is heard, beneath the sketches of each generation.

Will, who has worked on the farm for nearly seven decades, knows about life and death. He knows he's not forever for this world. The Harris men often go out with dementia. It's not pretty.

## River of Life

Will believes in genetic predisposition and he hates that this could be his future. He hates it for his family. But if there's one thing that's clear on a farm, it's that death, however it comes, is part of life.

"Everything that lives dies, and nothing that lives wants to die," Will told me during a visit to the compost pile the size of a football field and stacked taller than any building in Bluffton. But, he continued, "In a healthy ecosystem, nothing stays dead long. It'll be eaten and it'll live again." White Oak Pastures uses every bit of the cow, and what can't be used gets composted.

When he dies, Will wants to be composted, too. He wants to be laid naked—maybe a sheet over him, but he's not modest—on a five-by-ten-foot slab of steel covered with layers of organic matter—peanut shells, wood chips, maybe some scraps from the kitchen, or a little horse manure from the barn. Once a year he wants another layer added.

Will wants to become the thing he's worked so long to save and build. Carbon. Nutrients. Minerals.

Will wants to be soil.

# Part III: Regenerate

# Chapter 8
# Back to the Land

# Small Farms, USA

**Regenerative Agriculture Thrives on Diversity**

So far, most of the American farmers I've met have been on the land for generations. They've benefited from the inheritance of acreage and knowledge. In this country, first-generation farmers are rare, and young ones (under the age of thirty-five) account for less than 7 percent of the total.

But they're not as rare as farmers of color. I was learning that diversity is at the heart of many of the farms I visited, but it's clearly not a value for the US farming industry as a whole: Only a tiny fraction—fewer than 2 percent—of American farmers are people of color. And yet, they are some of the movement's most powerful teachers and elders.

I wonder if a regenerative farm future is even possible without diverse farmers of all ages, many of whom hold ideals that I've come to realize are rooted in cultures and practices outside of the industrial agriculture promoted by the West.

Of course, there are a lot of obstacles in their way—including histories of colonialism and genocide, and today, still, racism and wealth inequality, as well as the sheer cost of land.

Plus, do young people actually want to farm? That question kept coming up in conversations with farmers from Georgia to Ohio to Hawai'i.

Hand-picked turnips from Howard Allen's Faithfull Farms in North Carolina, heading for the farmers market, where he will feed his community. FAITHFULL FARMS

And if they do, is it even possible for them to afford land and the necessary equipment to get started?

Then if they are successful, what will happen to their land when they are gone? With a little help and attention to the cycles of life and death, soil regenerates, forests regenerate—but often the generation part of regeneration is left out when we're talking about the people involved. No one outlives their land. So who will live on it next and will the knowledge of that land be passed on?

I traveled to three farms across the United States with these questions in mind. Each of these farmers found unique paths to the land and to seeing it and their opportunities on it as gifts to be shared.

•

### Faithfull Farms (Chapel Hill, North Carolina)

It was a cool October morning at the farmers market in Carrboro, tucked in among the forests and abundant parks of central North Carolina. Years ago, I worked in Carrboro at a locally owned retailer that sold backpacks and tents, which is to say I got really good at hacky sack. It's the kind of town where a spontaneous drum circle might form. I knowingly ate my first piece of organic fruit, an orange, in Carrboro. But I'd never visited this market before.

Howard Allen stood before his fresh-picked offerings—a lettuce mix, pea shoots, microgreens, and muscadine grapes—displayed in homemade wooden boxes sitting on a tablecloth patterned with colorful mandalas. He'd likely done the picking himself. He'd certainly done the planting and growing. And now he was doing the critical last step: connecting with people in his community.

"Do you have any of those little hot peppers left?" a woman in a hand-knit red beanie asked.

She was pickling peppers and had screwed up her first batch. Howard, who's worked as a chef and baker in New York City and here in North Carolina, empathized but replied, "All we have on the farm are jalapeños, but I didn't bring any this week."

"That's really all I need," the woman said.

"I can bring you some this afternoon," Howard offered. "I'll give you a text after the market and then I'll drop them off."

He didn't need to ask for her number. He didn't need to ask where she lived. Howard knew the people he fed. He asked a shopper about their vacation in Greece. Another one stopped by to say hi after having been gone for a few weeks following an emergency appendectomy. He invited her out to the farm to breathe in some fresh air, pick some vegetables, and continue to heal.

Another shopper introduced her sister, visiting from California. She wanted her to try Howard's muscadine grapes.

He reached into a container and selected a grape larger and rounder than what you find in the supermarket. The grapes were green, black, and red. Some looked like large hand-painted marbles, others like tomatoes, and one even resembled a tiny watermelon.

"These are native American grapes that have been grown here for centuries," he explained. "They're good for jams, jellies, and pies. They're definitely a Southern thing."

The woman plopped one in her mouth, and Howard told her she could swallow the seeds or spit them out.

"I eat them like candy," the local sister said.

They bought grapes and a salad mix with a tap of their credit card on Howard's reader, the financial transaction of $13 barely registering when compared to the ongoing social interaction.

And the people he fed knew Howard.

They asked about his son, who sometimes runs the Faithfull Farms stand on his own. Howard is thoughtful and quiet, but his ten-year-old, outgoing and chatty, is a natural salesman.

Howard started Faithfull Farms for his children. He wanted them to have a connection with the soil like the one he had when he was a free-range kid in Jamaica, picking his snacks from trees. He had worked alongside his parents, who grew some of their own food using practices that didn't have a fancy name, but resembled permaculture: When they harvested banana plants, they'd chop the leaves and use them as mulch to keep the soil moist and suppress unwanted vegetation. No one used chemicals. He wanted to pass on what they'd taught.

Howard started farming for his family on one-tenth of an acre in his backyard a few miles from where we stood. Since then, "Things have kind of gotten out of hand," Howard said, chuckling, and perhaps considering the complexity of crop rotations, variety, and planning that goes into running his current system, or how he left his job to take a 95 percent pay cut to farm full time.

As he began to expand, he looked to books and videos featuring intensive vegetable operations, which grow a lot in small spaces

by focusing on soil health. He took in what the big names—like Eliot Coleman, who produces year-round vegetables on an organic farm in Maine, and Jean-Martin Fortier, who famously grosses $100,000 of produce per acre on his operation in Quebec—had to say. Their success and the profitability of other small-scale vegetable farmers demonstrates a path forward for new farmers.

People who are locked out of conventional agriculture because of the cost of land, equipment, and agrochemicals often include three demographics—immigrants, young and first-generation farmers, and people of color. Howard belongs to all three.

Yet, these groups, based on necessity and culture, are often the foundation of regenerative agriculture. J.I. Rodale and his son Robert, who published Rodale's organic farming and

Howard and his family. Only 2 percent of US farmers are people of color. Howard left his job as a chef, taking a 95 percent pay cut to start farming on one-tenth of an acre. JOHN MICHAEL SIMPSON / CHAPEL HILL MAGAZINE

gardening magazine, were among the leaders of the organic movement Carmen Fernholz was part of, and are often cited as the first to use the term "regenerative agriculture." But these White men didn't invent the methods they promoted through their magazine and later the Rodale Institute. They were inspired by Sir Albert Howard, who learned from traditional farmers in India. And permaculture as well as Allan Savory's holistic management were influenced by Indigenous practices.

Dr. George Washington Carver comes up, as he should, in most conversations about Black farmers. Born into slavery in the 1860s and raised by his former enslavers after abolition, he made a profound impact on the history of American agriculture, especially of the non-chemical variety. The first Black student to attend the school now known as Iowa State University, he became a professor at the Tuskegee Institute and one of the first scientists to reject chemical fertilizer and the industrial monocultures of his time, which treated soil like a plant-growing machine. His work focused on cover crops, crop rotations, and compost, each of which is now seen as an important part of growing regeneratively. In fact, Liz Carlisle, a professor in the Environmental Studies program at UC Santa Barbara and author of *Healing Grounds: Climate, Justice, and the Deep Roots of Regenerative Agriculture*, writes that Carver is "arguably the first US scientist to ... call for regenerative organic approaches."

As industrially produced, synthetic inputs began to be used on farms, he recognized that they were harmful to the soil and also out of reach for Black farmers. So, as Carlisle explains, if he could show Black farmers "what to look for in the natural world... he could show them how to win their economic independence...."

## Back to the Land

A century before Will Harris, Carver encouraged people to see the cycles of nature by interacting with them. "Reading about nature is fine," Carver said, "but if a person walks in the woods and listens carefully, he can learn more than what is in books...." He encouraged a deeper connection and conversation with other people, soil, and all life, stating that "if you love it enough, anything will talk to you."

In 1910, Black farmers owned 14 percent of the total land farmed in the United States. That number has plummeted, down to less than 1 percent. Leah Penniman, a Black farmer and educator, writes about what led to this in her book *Farming While Black*:

> Our Black ancestors were forced, tricked, and scared off land until 6.5 million of them migrated to the urban North in the largest migration in US history. This was no accident. Just as the US government sanctioned the slaughter of buffalo to drive Native Americans off their land, so did the United States Department of Agriculture and the Federal Housing Administration deny access to farm credit and other resources to any Black person who joined the NAACP, registered to vote, or signed any petition pertaining to civil rights. When Carver's methods helped Black farmers be successful enough to pay off their debts, their white landlords responded by beating them almost to death, burning down their houses, and driving them off their land.

Howard is well aware of this history and is committed to slow growth independent of loans or grants from the USDA. He was even skeptical of the loan forgiveness program meant to help Black farmers during the COVID pandemic. He thought it was

the agency's effort to throw Black farmers a bone. Even though the USDA admits past wrongdoings, I can see why Howard is skeptical of the organization's commitment to Black farmers. In 2022 the USDA granted direct loans to only 36 percent of Black applicants. Compare that to the 72 percent of White applicants who were successful.

"You know, there's so many injustices that are systemic," Howard said, his words coming faster. "There's always some political people using these things to say, 'Hey, we did something for Black farmers.... Vote for us!' But to me, it's a slap in the face. A lot of farmers who are on the forefront of these fights [seeking assistance from such programs] are perpetuating the system. They want an equal share of a system that's

Groundbreaking agricultural scientist and inventor Dr. George Washington Carver, born into slavery, promoted regenerative and organic practices more than a century ago. BETTMANN / GETTY IMAGES

already unjust. The whole agriculture system was designed for White farmers."

Howard told me about how the USDA has created "hog cycles" of boom and bust throughout its history by encouraging farmers to get into hogs and offering loans to support it. Soon, the market floods, the price of hogs drops, and farmers lose or sell their farms. This especially hits Black farmers because they have less access to resources and programs that could help them weather the price drops.

Howard doesn't have much sympathy for them. "You spent all the money buying, borrowing, just like the big guys, to do hog production and suddenly you're blaming the USDA," he said. "But you're just doing something to make money, and you weren't looking at this from a holistic perspective."

When Howard expanded, he didn't borrow money. He borrowed a corner of his church's land tucked next to some woods and began growing food for the members of the church. That evolved into a CSA that provided vegetables for twenty families, including some in need. He added hoop houses and started selling year-round at farmers markets and to restaurants. He looked at it as a ministry and welcomed volunteers from the church and the local universities. Next he expanded onto an older neighbor's land. Howard's operation, which now totaled three-quarters of an acre, is spread across four different pieces of land near his home, and he talked knowledgably about each location's intricacies and microclimates.

Later when I visited Howard's farm at the church, I met Effua Sosoo, who had recently earned a PhD from the University of North Carolina. Effua is a clinical psychologist and researches the impact of internalized racism. She told me that it's challenging to get Black people to engage with farming. "They carry a lot of trauma associated with the land," she said. "If you mention farming as a career to someone, they can see it as an insult."

Makes sense. When the first iteration of industrial agriculture in North America was fueled by the enslavement of your ancestors, I see how it can be traumatic to stand in the sun weeding and turning the soil.

Leah Penniman and her family started Soul Fire Farm in Pennsylvania on seventy-five acres. She speaks to Effua's point in her book: "While farming was initially healing for me, for many African heritage people, it is triggering and re-traumatizing. Almost without exception, when I ask Black visitors to Soul Fire Farm what they first associate with farming, they respond 'slavery' or 'plantation.' As fourth-generation farmer Chris Bolden-Newsome says, 'The field was the scene of the crime.'"

Penniman, who works to help others heal from that crime, writes powerfully about the trauma it has caused. She continues, "Hundreds of years of enslavement have devastated our sacred connection to land and overshadowed thousands of years of our noble, autonomous farming history. Many of us have confused the terror our ancestors experienced on land with the land herself, naming her the oppressor and running toward paved streets without looking back. We do not stoop, sweat, harvest, or even get dirty, because we imagine that would revert us to bondage."

But she also expresses hope: "And yet we are keenly aware that something is missing, that a gap exists where once there was connection. This generation of Black people is becoming known as the 'returning generation' of agrarian people. Our grandparents

## Back to the Land

fled the red clays of Georgia, and we are now cautiously working to make sense of a reconciliation with land. We somehow know that without the land, we cannot return to freedom."

Howard is part of the "returning generation." He taught Effua how to plant, weed, and cultivate. Now she grows microgreens in her apartment and volunteers regularly on the farm.

"Howard makes you want to grow and eat vegetables," she said.

While Howard acknowledges Black Americans' greater historical, economical, and psychological barriers keep them from farming, he believes that all of us have become disconnected from the land. He works to bring everyone back. He welcomes volunteers—there's always something that needs to be done—hosts corporate team-building activities on his farm, and imagines a future where people can stay in a guesthouse there and walk the edible nature trail he has planned and eat the mushrooms he's learning to grow.

Howard's upsell is always an invite to visit. I took him up on it, and after a tour offered to help out, along with several other volunteers. We put down wood chips in the paths between the beds in the high tunnels and we started some radishes, but I'm not sure we were all that helpful. Howard was more efficient and proficient at every task. In the time it took him to show us, I think he could've done all the work himself.

Howard makes a profit. He supports his family and his community on less than one acre. He knows his numbers, and probably could tell me to the foot how much drip irrigation tubing he has. He and his wife, Ronniqua, have a business plan and a small, devoted staff. But what started as a ministry, remains a ministry.

Before I left the farmers market, a woman approached whom Howard didn't know—I could tell because he didn't immediately greet her by name.

"I don't live here," she said. "But farmers markets are my favorite place. Not to be disrespectful to anyone else's relationship to their spirituality, but I call it church."

"And we have a wonderful congregation," Howard replied.

"Can you tell me about the mission of your farm? I'm wondering what it means to be 'beyond organic,'" the visitor asked, pointing to the Faithfull Farms banner that read: "Growing beyond organic. Steward of the Land. Serving Our Community."

"Our mission at the farm is simple," Howard said. "It is to connect people with the land in a tangible way. We do that through food and actually getting people on the land, and beyond food, we provide a space where the community can become one with the land."

"That's beautiful," she said.

A little boy, maybe hers—there were a lot of free-range kids there—marched up to Howard and asked what he could buy with a dollar.

"What do you like?" Howard asked.

The boy pointed and said, "I love pea shoots."

Howard laughed. The boy handed Howard a dollar and Howard handed him $4 worth of pea shoots. An investment in the future. The next time Howard will probably invite him to the farm to plant some peas.

Howard practices intensive vegetable production, allowing him to support his family by farming on a small amount of land. He grew up in Jamaica and learned to farm from his parents. FAITHFULL FARMS

**Meadowlark Organics (Ridgeway, Wisconsin)**

Back when Paul Bickford was a fourth-generation grain and dairy farmer, he was all about investing in the future. He didn't want to be the last to work his family's land. When a change needed to be made, he made it. In 1978 he was one of the first farmers in his area to take his herd off grass and bring them inside onto concrete, sleeping, eating, pooping, and being milked all under one roof in the name of efficiency and profit. He brought the food to the cows and took the manure away. When milk prices dropped in the '90s and his costs rose, he made another change. In 1992, he put the herd back on pasture and began to sell grass-fed milk.

The university extension agents thought he was crazy. His neighbors thought, "Boy, he's really gone off the deep end now. He's planting fence posts." His own dad, his business partner, agreed with them, so Paul bought him out.

Paul and his wife started to rotationally graze their herd of three hundred cows across their roughly thousand-acre farm. Outside the confined model, milk production goes down, but so do the costs of feed and veterinary care, which dropped from $86 to $5 per cow. Turns out that fresh air and sunshine was a lot healthier. Paul's costs were down, and his profits were up.

But there was something wrong with the cows. Sometimes when they were in the milking parlor, they were reluctant to drink, and when they peed they would start kicking, acting possessed. He thought the problem could be that the cows were being shocked by stray current due to how the electrical grid was constructed. He knew this had affected other dairy operations, causing reproductive problems and lower milk production.

## Back to the Land

Paul and his wife sued the electric company and won. Good news. Similar cases had ended with millions of dollars in damages going to farmers, so they were hopeful. But in their case the jury awarded them exactly $0. Nothing at all. Bad news.

The losses and changes began to pile up. He sold his cows, sold his house, and plowed his pastures under. He and his wife got divorced.

In the middle of all this, his father was diagnosed with Parkinson's disease. Paul knew that Parkinson's, along with many types of cancer, was more common in farmers who applied pesticides, so he made the decision to stop using chemicals on his farm and planted organic grain, mostly soybeans and corn he'd sell for animal feed.

Paul was getting concerned about his lineage, since none of his children wanted to take over the farm. He had learned so much from his mistakes and bad luck. Who would he pass the knowledge on to? Who would continue to improve the land and the farm?

As a kid, Paul's son Levi had been diagnosed with a nonverbal learning disorder, which later was identified as autism. He was a valued worker on the farm but didn't have the skillset or interest to manage the farm, especially given the complexity of farming a diverse rotation of organic crops. If Paul just sold, what would Levi do? What would Paul do?

Paul needed a partner, someone who could take the farm into the future. As the years went by and he hadn't found anyone, he turned to an unlikely place in 2015: Craigslist. Paul's ad read:

> *I am seeking a forward-thinking individual or couple to join my 950-acre organic farming operation to assist in all facets of growing feed crops and to assist*

> *in marketing of corn, soybeans, small grains and hay....*
> *Ethics and trust are a cornerstone of organic farming*
> *and are important to my operation. I want to share my*
> *forty years of farm experience with someone who is*
> *willing to work to improve my farm.*

Time passed, no reasonable candidates emerged, and things were getting tougher. Paul was feeling the pinch in the marketplace as the price for organic corn and soybeans had been pushed lower by imports from Turkey and Romania. Somehow organically grown grains could be shipped across oceans and sold for less than the ones Paul produced. And some of what he was competing against wasn't really organic. On at least one occasion, a six-hundred-foot cargo ship of conventional soybeans from Turkey magically "earned" the USDA organic certification by the time it landed in the US port. The exporter had falsified records. Paul was among the growing number of certified organic farmers becoming disillusioned with the USDA Organic label, and knew that if he didn't find someone with ideas and energy, the next change he'd have to make might be his last.

Enter Halee and John Wepking.

•

Halee, a woman in her early thirties wearing purple horn-rimmed glasses, a flat-billed trucker's cap, and a maroon shirt with the word "Grainiac" written on it, greeted me in front of a not-quite-finished mill.

Against all odds, and to the likely surprise of every guidance counselor she ever had, Halee, who grew up in central Arizona and has a degree in modern dance, is a farmer.

**Back to the Land**

"This is still fifty percent Paul's farm," Halee said, describing the partnership she and her husband, John, have with Paul, now in his late sixties.

Halee and John met as chefs working at one of those one-word-named restaurants in a trendy neighborhood in New York City. She said its name, Prune, as if there were any chance I might recognize it. My hometown didn't even have a Chipotle.

Becoming increasingly interested in where their food came from, they left the city in 2014 to join John's family's grass-fed beef operation in Lancaster, Wisconsin. John thought he might operate it with his dad, but the other relatives involved were of another persuasion, and before long John and Halee knew they weren't going to have the life they had imagined on the family farm. It's an all-too-common situation. One of my Ohio Farmer friends just went through this with his family. It was horrible. Relationships strained. Lawyers involved. Christmas never the same.

"One day I was raking hay," Halee said, "and I was very pregnant, so I went home early. I got on Craigslist and typed 'organic' in the job search. And I found a bunch of 'pick weeds on our CSA farm' ads and then I found Paul's post. I was like, 'Holy shit!'"

It was 2015, and she and John were the first to make it through Paul's phone screening to the farm-visit portion of his interview process. It didn't take long before they sealed their partnership with a handshake and agreed to work out the details as they went. Paul took the insurance money from a fire, along with a loan, and made some improvements.

The farm is beautiful. Hills rising to the north and dropping to the south, looking like those random arcs that flow into one another

in a kid's first landscape drawing. The area, which had been mysteriously bypassed by the last glacier some ten thousand years ago, is known as The Driftless. It starts just south of this farm and runs north of Mark Shepard's farm sixty miles away in Viola. The farm was divided by a stream and a highway, and was like a patchwork quilt of colors and textures as each plot grew something different.

In a few long, slow strides Paul made his way over to Halee and me. He had things to do. I could almost see the list in his mind, but he pushed it all aside to meet me.

I told Paul how much I was enjoying the landscape. "I'm from a part of Indiana where it's nothing but flat fields. So I see a hill and I'm like, 'Wow!'" I said.

"Yep..." Paul said, and I thought he was going to leave it at that, a man of few words, but then he continued. "I grew up on a flat farm. It takes longer to farm on a hill. I guess you get more corn

rows or something." Paul seemed like a guy who throws in an "or something" to be humble, and who probably actually knows the exact number of additional rows per acre based on particular slopes, and who'd love to tell you about it if you asked.

Instead, I told him about my journey to understand regenerative agriculture.

"The key point of regenerative in my mind is right here," he said, pointing at Halee. Like I said, he's a man of few words.

LEFT: Paul Bickford was so desperate to find someone to pass his knowledge and land onto that he placed an ad on Craigslist: "I want to share my forty years of farm experience with someone who is willing to work to improve my farm." CLIFF RITCHEY

ABOVE: Halee and John Wepking had hoped to join John's family's farm after working as chefs in New York City. When that didn't work out, they looked elsewhere. They were so desperate to find a way onto the land that they answered Paul's ad. CLIFF RITCHEY

Halee and John are part of a small uptick in the number of American farmers under the age of thirty-five. Yet for every farmer under that age there are six over the age of sixty-five.

When they'd first met, Paul told the couple, "If you come to work with me, I expect you to tell me what I'm doing wrong and bring new ideas." He was ready to make more changes.

Halee and John were ready, too. They wanted to grow food for people, not for animals. In the spring they'd plant oats and barley, which they'd use to make hay, and also alfalfa. They'd till under the alfalfa as a green manure, a natural fertilizer, and then plant corn. Halee and John added a mix of small grains, such as wheat, spelt, and rye. And they wanted to build a mill, which would allow them to process their own grains to sell directly to bakers and consumers, and to work outside the consolidation that Will Harris talked about.

Even with his new partners, Paul seemed to be working as hard as ever. "Are you exhausted by all the work or do you still enjoy it?" I asked Paul, who planned to work at least another five to ten years on the farm.

"I'm exhausted, and I enjoy it," he said. But he acknowledged that his many years of making changes to survive were now behind him. "I'm content," he added. "It's pretty much their show now."

- •

A short drive away, John sat in a cab of what looked to me like a shrunken antique combine, maneuvering back and forth across a field. The machine, a swather, cut the wheat and windrowed it for drying.

## Back to the Land

Halee and I stood on a patch of field that had already been cut. We were joined by my friend Cliff Ritchey and his wife, Julie, who were taking photos for the Midwest road trip portion of my regenerative journey. Cliff had also traveled with me to Colombia to learn from the Arhuaco. I wasn't sure if he kept coming on trips because he enjoyed them or he was concerned about me. *Would Kelsey even remember to eat?* So Cliff and Julie packed a cooler full of food, made sure I drank enough water, and took photos.

When John reached our side of the field near the road, he cut the engine and joined us.

"It's a beautiful day," John said, and then gave us the weather forecast for the next seventy-two hours. He rubbed his tightly trimmed beard, which didn't have a single gray hair, and then pushed his horn-rimmed glasses back up his sweaty nose.

He explained what we were seeing in the field. There was spring wheat, cranberry beans (which apparently are a thing that exists), and some black beans that had enough weeds around them that I would have been concerned if I didn't know better. By now, I wasn't surprised to learn that Halee, John, and Paul didn't concern themselves with a few weeds. Like many of the other farms I visited on this journey, their crops thrived in good soil strengthened by a diversity of plants in fields fairly undisturbed by tilling.

I tried to imagine John and Halee walking down the streets of Manhattan to the restaurant where they met. Nothing in Manhattan is measured in acres. One hundred people-less acres stretched out behind the young farm couple. If I could snap my fingers and magically apply the population density of Manhattan to this land, we'd suddenly be joined by a crowd of

Paul harvesting organic wheat next to a strip of brush where John often forages for mushrooms. Halee's dad referred to Paul, with his impressive work ethic, as a "chip off the old Midwestern block." CLIFF RITCHEY

twenty-five thousand. The land, equipment, and the farmers themselves looked different than what I'm used to back home, but then there's not much about this situation that's typical.

"A farm like this isn't your usual entry point," John said. "We bought half the shares. We all draw a salary from the farm. That's nice because you can't really count hours in farm work."

As they work their way into full ownership, they have promised Paul to always have a place on the farm for Levi to work.

And they've created a real home there for their own family, including Henry, who was born in 2019, and Lyda, two years younger.

"If all we did was work to feed ourselves, we could," Halee said. "There's so much out there."

But of course they have bigger goals.

John walked through the entire farm plan, the whole six-to-eight-year rotation, including caveats for expected unexpected events. He talked about spring and fall plantings, about frost seeding, heritage wheats, nitrogen, fertility, buckwheat, and spelt.

Unlike the farmers back home, he never mentioned yield.

I thought about my friend with his corn prescription from the seed company and his GPS, which told his planter and sprayer what to do on every inch of his farm. John's schematic was in his head, where it was created, and I'm pretty sure he could hand draw it from memory.

There are so many more variables in Halee and John's plan. They're thinking about conserving and protecting the land, so it will be healthy and productive if their own kids take over the farm.

# Back to the Land

"To farm with a conservation ethic," John said, "and at this scale, we have to be a lot more creative in the crops we grow and create the diversity in the system we want."

Something else they wanted—to feed people—is bringing them closer to the land and their community. "Paul has these old family pictures with Bickford big milk bottles," Halee said. "They were direct-marketing the milk. And that's not very different from what we're doing. So it's like getting back to that kind of commitment to the land and life, realizing that you have to take ownership of it through to consumer."

"Do you miss anything about the city?" I asked.

She paused and shook her head. "I feel so much more connected to this place. We're finally feeling the benefits of settling. Like the cherry tree we planted had cherries on it this year. Like those kinds of things. When you put down roots, it takes a while before you are in bloom. And it feels like that happened this year."

It gets better.

She continued, "It feels like with the friends we have made out here we have deeper, stronger friendships than the people we were friends with in New York. [Here] you rely on your community."

"We're lucky to be in the place where we are," John added. "There's a lot of pretty vibrant small towns and a lot of progressive rural communities."

•

"Hey Henry," Halee said to her four-year-old son as we sat behind their house overlooking the farm. "Kelsey has been to Africa."

"And I've seen lions," I told Henry. His eyes got big.

"Did you see water buffalo?!" Henry said in a voice that jumped pitches like a songbird's.

"I'm not sure if I did."

"Water buffalo have those curved horns," Henry said. "Did you know that water buffalo are predators to lions?"

"Does a water buffalo eat a lion?" I asked.

"No! Water buffalo are herbivores and lions are carnivores!" he said, pronouncing each word precisely, adorably, and with an engaging enthusiasm that made me think that a fine replacement for David Attenborough's revered authority as a wildlife film narrator would be that of an obsessed four-year-old boy named Henry.

Each of us took turns pitching a Wiffle ball to Nature Boy, who'd crank it into the garden or beyond the fence that separated the pasture from the yard, and then he'd flip his bat and do a home-run trot around imaginary bases.

When dinner was ready, we sat at a table beneath an oak tree that must've been two hundred years old. It had seen farmers come and go. John bounced Lyda on his knee as she watched her expressive brother give us all lessons on life and nature. Our plates were empty and our bellies were full of polenta, veggies, bread, and chorizo. Henry especially loved the chorizo, which came from a neighbor. Everything was local. The veggies were from the garden a short walk downhill from the house.

The sun dropped beneath the thin layer of clouds and the sky shifted with pinks and reds.

## Back to the Land

Halee's dad, who was living in his camper next to the house, had a bushy mustache worthy of a Civil War general. He had worked for years in construction in Arizona and besides helping Halee and John refurbish their home, he picked up some jobs here and there. He appreciated the farm's location for its access to good trout-fishing streams and the abundance of other foods.

"John went down to deal with the cattle last night," Halee's dad said. "Came back with no shirt on. He's got this shirt-full of oyster mushrooms."

Halee looked around. Who knew who'd first lived here, what their lives had been like—but it was clear they hadn't put down the kind of roots she had. "Funny thing about this farmstead," she said, "is there were no lilac bushes, no asparagus patch. Like, what were people even doing?"

Julie agreed. What were they thinking? Past-Julie and past-Cliff had planted food for future-Julie and future-Cliff to enjoy back home at their own place. Mint for mojitos. Strawberries for ice cream. Asparagus for butter. When Cliff talked about his herbs and spices, he said, "They are super precious to us."

Julie, Halee, and Cliff marveled that people wouldn't choose to live like them. I nodded, realizing I had been the people they were talking about. Until recently, I didn't have the patience for or interest in planting a garden or waiting years for a tree to produce food. I made a note: Plant an asparagus patch, a lilac bush, a patch of raspberries, and a cherry tree.

From the top of the hill, we could see the Dodgeville Walmart seven miles away.

"Do you hear the combine?" John asked Henry.

"Yeah!" Henry said, "it was there and then over there and then over there!"

Paul was harvesting a rolling hill of wheat backlit by the sunset.

"He's a chip off the old Midwestern block," Halee's dad said. "'Oh, it's a holiday.... Guess I'll work only twelve hours today.'"

Paul disappeared down the hill and then we couldn't hear the combine anymore. At the time, no one could've imagined it, but it wouldn't be long until he would die in an on-farm accident that would leave the responsibilities of the farm solely on the Wepkings.

Halee and John would write about their loss in their newsletter: "There is no one word that can describe who Paul was to us. He was

Helping the Wepkings "rehome this hill" by joining them for a delicious dinner and beers beneath the giant oak tree behind their house. Cliff and Halee's dad played guitar and sang songs until Paul was done harvesting and the lightning bugs were out.  CLIFF RITCHEY a mentor, a friend, and a father figure, though he never liked to be called our boss. He was an honorary grandpa, aka Bumpa, who took our kids for tractor rides and always had a cookie or an orange to give them. To say he will be missed doesn't begin to capture the loss we feel."

But that evening beneath the oak tree, time stood still. Halee's dad pulled out a guitar. Cliff grabbed his. They started plucking and tuning and feeling each other out. Cliff's eyes lit up. Halee's dad could play.

Henry had successfully lobbied to stay up to see the lightning bugs, which started to rise from the pasture. He gave chase.

Cliff let Halee's dad have the first song. He played "Metropolis" by Anthony Smith. It's about a guy who leaves a dying small town with a population of 404 for the big city. He lives in a high-rise with a "concrete yard," meets a girl, falls in love. They are expecting their first child and the mother-to-be says they are going to need more space and he knows just the place where air and water are free. They move back to his hometown, and the population increases to 407.

As the final chords faded, Halee and John took it all in, content in their choice "to re-home this hill" with their young family.

- 

### Tree-Range Farms (Northfield, Minnesota, via Guatemala)

Reginaldo (Regi) Haslett-Marroquin and his father and brother were heading through the jungle near their Guatemalan village with a purpose. His father wore a hand-pumped backpack sprayer filled with the chemicals a visiting agronomist had sold him, hoping they would finally kill the chickweed taking over their field. After years of battling it with machetes and fire, they were ready for war.

Suddenly, a venomous fer-de-lance descended from a tree and wrapped around his father's arm. It was the same kind of snake that had killed a family member a few months previously. He froze. While Regi stood there, unable to move, his brother raised his slingshot, aimed, and dropped the snake.

Shaken, they took a moment but then continued, and Regi's father went on to pump and spray a different kind of death. It was the first herbicide the family would ever use on their fields. It made Regi's head spin and he could still taste it hours later. He thought there was no way it would work, but when they returned to the fields the following week, the chickweed was dead.

Regi, who was a teenager at the time, never forgot how amazed and impressed he was. "I need to know what happened here," he told himself, vowing to go to school to become an agronomist.

He also never forgot his father's reaction. "Death," his father responded. "Death came to this field, and death is all that we'll harvest from it. That chemical I sprayed, it's still here, in the soil, on the corn, in the air. How can we eat food that has grown on death?"

But they were desperate. The family had lost crops before and had experienced hunger. Even so, his father vowed to never use chemicals again.

Many years later, Regi would take these memories, his ancestral knowledge—his roots were deep in Maya culture—and his agronomy training with him as he began to create a new agricultural system, one that would help farmers out of poverty. I met Regi far from the jungle, far from Guatemala, on his 1.9-acre homestead near Northfield, Minnesota, where he has devised a chicken-based farming system to revolutionize agriculture.

In 2022, Acres USA presented Regi with its Eco-Ag Achievement Award. Past winners included Gabe Brown and anti–Green Revolution activist Vandana Shiva. At the event, he was introduced as "a passionate visionary whose dedication to pushing for smart, regenerative change is unmatched" and that he "has been a strong leader in the movement for developing innovative poultry-centered regenerative agriculture systems around the world."

I was excited to meet him. His head of salt-and-pepper hair came to my shoulder. He wore a black T-shirt tucked into khaki pants, and carried a thermos. I had interrupted his writing retreat in which he was putting together an online curriculum so he could teach his methods more widely.

Nothing about the farm seemed that remarkable or visionary at first glance. We started at his family-sized garden, which was about as chaotic as mine. His rows of corn were spaced haphazardly and a few stalks rose at odd angles. But as he explained how it worked, it was soon obvious that a lot more thought went into his chaos than into mine. Clouds of cilantro and sage and mountains of beans surrounded the corn and outcompeted any unwanted vegetation. The herbs' scents confused would-be pests. The sweet corn, which was about the same height as Regi himself, had tasseled and provided shade for hazelnut saplings that would become a critical part of his system.

"These are beans from Guatemala," Regi told me, pulling back velvety leaves to reveal finger-length pods. He doesn't sell anything from his garden, but he grows enough to feed his family for the entire year. Food sovereignty—people being able to control their own food supply, policies, and distribution—is important to him as a father but also as a citizen. In Guatemala and in the United States he has witnessed firsthand how unjust and exploitative corporate-controlled, industrial agriculture can be, and emphasizes the importance of people being able to feed themselves.

"There are millions of immigrants who work in the food system, from farms to factories in the United States," Regi said, tracing the roots of an industry that exploits immigrant labor back to one that enslaved and built wealth upon the labor of Africans. "It's the unseen history of modern colonialism and slavery. Now we've got this inconceivable amount of wealth in the hands of a few, and an inconceivable amount of work and value produced by the masses. That degenerative system requires change."

The method he's developed, by contrast, is surely regenerative, in several senses of the word. It rebuilds things that are broken and destructive, starting with the people involved. His process, he says, "allows anybody with interest to enter the system, even if they are an immigrant who doesn't have land or savings." He calls his brand Tree-Range Farms, and it allows "hard-working individuals who understand the relationship with nature" to make a new start, no matter who and where they are. If you have access to a little land, you too can be a Tree-Range farmer.

The chickens improve the soil and the lives of the farmers. Soil and people, humus and humans, earth and humanity, forever interlinked. Degenerate the soil and the human community also degenerates. But renew it? And, well you get the idea. It's a

relationship that each of us can see play out in a garden, as Regi did as a boy.

"Cilantro is my favorite herb," Regi said. He doesn't have to plant it. He lets it go to seed and it plants itself.

His love for cilantro has deep roots. Back in Guatemala, he began applying himself in school after he'd made his mind up to become an agronomist. In his area, most families could not afford for their kids to continue past sixth grade. To help cover his costs and support his family, Regi grew the herb—"the best in town"—in his family's garden and sold it in the market, making two trips there on his bike before school. He also mucked out cattle stalls for a neighbor. While other kids expected to be paid in cash for that job, Regi accepted his payment in manure, which was the secret to his cilantro business.

Regi Haslett-Marroquin, originally from Guatemala, built a chicken-farming model that can provide immigrants and the poor meaningful and profitable work. WIL CROMBIE

It had become a struggle for small farmers in Guatemala to make a living. Regi grew up in the 1980s in the middle of a thirty-six-year civil war. The previous government had offered more support for small farmers and Indigenous people, but the new US-backed administration opened the agriculture sector to larger corporate influences. The anti-government guerrilla forces were often made up of farmers, so farmers became common military targets. Guerrillas sometimes attacked farmers who weren't aligned with them. Being a farmer during a civil war was to observe that corn grew taller and greener atop mass graves.

Against the odds, Regi was accepted into ag school when the time came, but he couldn't afford the required boots. A friend of his brother-in-law, who was a soldier, gave him a pair. But then as he was on his way to the city to start school, government forces stopped the bus and searched luggage. When they found the boots, they accused Regi of killing a soldier and stealing them. They kicked him, hurled insults, and gave him to the count of ten to run. He beelined for the cover of a cornfield and dove to the ground as they opened fire. Eventually, he made his way to a farm, where a family fed him and gave him the money he needed to finish his trip.

At ag school, Regi learned about the 'cides and artificial fertilizers. Graduating as one of the top students, he was hired by an agrochemical company and became the agronomist visiting families, selling death as a solution. He had money, could afford nice boots and a car, and he convinced himself that the chemical inputs would offer a short-term solution to poor families who needed one.

But something important was missing, and his father's vow to never use chemicals again rang in his ears. He wanted to

## Back to the Land

operate from a place of life and regeneration rather than of death and degeneration. So, after the trial period at his job ended and he was offered a contract, he quit. He started growing food at an orphanage where he met his wife, who was from Minnesota, where they eventually moved together.

Before I went to Northfield to meet Regi, I studied his memoir, *In the Shadow of Green Man: My Journey from Poverty and Hunger to Food Security and Hope*. I knew he wouldn't want to waste time standing surrounded by the smell of cilantro, telling me about his life in Guatemala and his journey to get here. He was more focused on what was ahead of him and us, the heart of his system—chickens.

Like Regi, chickens aren't native to Minnesota. They originated in the jungles of Southeast Asia. His flocks were behind a fence in the pasture, a few paces away. But this was no typical pasture. It actually sort of looked like an orchard, with hundreds of trees, mostly bushy hazelnuts—which Regi had purchased from Mark Shepard a decade ago—studding the area. Dozens of chickens crowded in their shade. There were other trees, too, including elderberry, from which Regi extracted juice to treat the birds and keep them healthy.

"I got to grab a picture of these guys. They look so comfortable," Regi said. It's true. This was no Southeast Asian jungle, but every inch of shade beneath the tree nearest us was taken up by a contented chicken. They love the protection the trees provide from the sun and from birds of prey.

Feeders hung from a simple wooden A-frame structure that would be moved around the pasture in order to spread the positive impact the chickens made on the soil with all their scratching, pecking, and pooping. Regi spends about forty-five minutes a day moving and refilling the feeders.

He has 715 birds on his pasture, which measures under an acre. If a farmer has 1.5 acres, Regi recommends a flock of fifteen hundred. The system also calls for a centrally located barn to house the birds every night.

Tree-Range farming, which Regi honed and teaches on this land, is designed to be accessible to people without large acreages. And it's focused on an animal, not plants, for a reason. Early on, he'd grown vegetables, but they were perishable and were always sold into a buyer's market. Everyone had tomatoes and cucumbers at the same time. He felt like the return was too little.

"Picking an animal was central," Regi said. He explained that animals speed up the flow of energy on a landscape. When there are no animals on the land, the flow of that energy is halted or reduced, requiring more work or expensive chemical inputs.

I'd ask him more about the "flow of energy" later.

Large ruminants were too costly and required too much space. He chose to focus on chickens because they fit into the diets of most religions and cultures and because they could be raised in a relatively short amount of time on a small amount of land. The Tree-Range system requires little investment up front.

Regi is so inspiring. If he were to host an infomercial promoting his method, I think he could convince anyone to become a producer. He certainly convinced me. As he described it, I imagined my front one acre—which currently houses nine chickens—filling up with a flock of hundreds of birds. Regi made me believe that even a farming doofus like me could be successful.

## Back to the Land

But a thought occurred to me: chicken manure is nasty. When it's sprayed on the big farms near my home, you can't go outside for at least a day. Chicken CAFOs, which house up to fifty thousand birds, give off the pungent aroma of ammonia and sewage, driving nearby homeowners to move away, often selling at a loss, if they can sell at all. Not everybody wants to be neighbors with, or in the case of my wife be married to, a chicken farmer.

"But what about the smell?" I asked.

Regi inhaled deeply through his nose. "Honestly," he said, "this is the worst it will ever get."

I barely noticed anything. The microorganisms and bacteria in healthy soil quickly break manure down before it begins to stink.

"So, if I took one acre and used your system, what kind of income could I expect?" I asked.

Regi looked up and away from the flock and started his calculations.

"Each chicken weighs an average of four pounds at sixty-five days, maybe seventy-five," he said. "In your area, you can probably raise easily four flocks per year." He goes silent. "How much does chicken sell for in your area?" he asked.

"We eat chicken every week, so I'm embarrassed to admit it, but I have no idea." Regi switched courses and worked with the numbers of his farm.

"On a full flock, you can gross more than $70,000. I'll send you a PDF that breaks it down. If I wanted to just hunker down and focus on this," he gestured to the mass chicken siesta beneath

the hazelnuts, "this place is a goldmine." That sounded unexpected, but promising.

"But," he continued, "we're not in the chicken business—we're in the business of quality of life. That's what we deliver. We've got a revolution here. And building a revolution requires some insurgency."

The US chicken industry has become incredibly centralized. Only four companies (you probably know the names) produce over half the chicken we eat. To revolutionize the industry, Regi started a nonprofit, the Regenerative Agriculture Alliance, which owns and operates the processing plants that he and other Tree-Range farmers use. Someday he hopes the alliance will also oversee their own hatchery. To him, it doesn't matter who owns

Regi sees his agricultural system as an act of decolonization, and believes that regeneration requires Indigenous ways of thinking.
LEIA VITA MARASOVICH

## Back to the Land

what as long as the farmers themselves have access and control over the system.

"If you are a new farmer or a farmer at a disadvantage, like me from an immigrant family who comes in without land," Regi explained, "you can work in our system and don't have to struggle." The Regenerative Agriculture Alliance was started in 2018. Just like a perennial system, time and patience are needed to allow the roots to grow and spread. As of 2023, there were six others in the alliance, including farmers originally from Mexico and El Salvador. They are all in Wisconsin and Minnesota, which puts them in proximity to the group's processing center.

In a small barn, Regi shows me the only tractor he'll ever need, a John Deere the size of a large lawn mower. He hopes that as his group of farmers grows, they will start an equipment co-operative from which they can each rent equipment to spread out the cost of its purchase and upkeep. In all, the chickens would support their ecosystems of trees and vegetables, but they'd also support a diverse ecosystem of twenty-some businesses controlled by the farmers, including the nonprofit, along with a processing center, hatchery, trucking service, and hazelnut-processing center. I had the feeling that if Regi started listing all the businesses, he would think of more to support those and the list would be endless. As he puts it, this idea of not wanting to own everything and not be controlled by a few individuals or corporations that do own everything requires that we "decolonize our minds," valuing quality of life and relationships over unhealthy pursuits, over profit and control, working together instead of working to dominate.

Modern history has been written and shaped by colonizers, those with power who've taken over and displaced landscapes and cultures. In Guatemala, the Spanish imposed their language, culture, and disease on the Maya in the sixteenth century. More recently, American companies, such as United Fruit, took over the land and exerted their influence on Guatemalan politics, agriculture, economy, and culture. Regi explains that decolonizing the mind means not imposing upon but working with and honoring and preserving the uniqueness of each plant, place, culture, and person, or otherwise just leaving them alone altogether.

He thinks that regenerative agriculture requires Indigenous ways of thinking. He saw my wheels turning as I thought about owning my own chicken farm, reducing chickens to pounds and dollars over time. He can speak that language, too, but in order to help me see things differently, he offered a memorable image: "Let's say that when you die, you volunteer to be chopped up and turned into mush and be put through a spectrometer for a chemical analysis. And then we did the same thing to the space here." He motioned toward the garden, chickens, and pasture. "Like this place, you are just energy."

"So we're just part of that energy? It flows through us and in us?" I asked and for a moment I felt like I was repeating questions to a Jedi in *Star Wars*, trying to learn about the Force.

"Yes," Regi said. "That's Indigenous thinking, seeing that energy, interacting, and relating with it."

Regenerative agriculture is working with that energy. That sounded like something the Arhuaco, Celestine, and Will Harris would agree with.

## Back to the Land

To Regi, it's fundamental to understand that what he and other regenerative farmers *don't* do is produce products. Instead, they act as stewards of a system that transforms non-edible forms of energy—sunlight, soil, air, compost—into edible forms. And since nature provides all those energy sources, regenerative agriculture essentially requires no inputs.

"If you look at what's wrong with the current system of extraction and colonization and degeneration," Regi explained, "it is the human factor—that need we're all born with to survive, to fight, to grab, to accumulate, to exploit, to extract. Nobody escapes that. Humans have been dispossessed, isolated, and pushed to the edges. All we can do is provide cheap labor to that system. The human factor basically has two ends to it. On one hand, you have the uncolonized regions of the world that were led and managed, governed and structured under an Indigenous way of thinking. Authority was earned, passed on, and revered. Indigenous people successfully balanced the human factor. On the opposite end, you have the colonizing system, which is the reason we have degenerated to the extent we have. The colonizing mindset unleashed the human factor and extracted, destroyed, and dominated."

Regi is concerned that the colonial mindset could negatively influence the movement toward actual "regenerative agriculture."

Of course, he's critical of the efforts of the large agrochemical companies to define and take over regenerative agriculture for their own purposes, but he's also disappointed with some of the genuine efforts of regenerative certifications. The certifying bodies haven't always consulted with Indigenous people and are too top-down for his liking. In his eyes, the organic, sustainable, and local movements have also been colonized by large corporations.

"Regenerative agriculture is becoming washed down and whitewashed, reduced and simplified," Regi said, echoing Will Harris's concern. "Sure, cover crops, thinking of the soil microbiome, perennials—it's all part of the process of regenerative agriculture. But regenerative isn't just about practices on the farm. It's about processes, about ownership and control. Agriculture is regenerative, or it is not."

I've been thinking about certifications for years, since I first visited the Fair Trade Certified coffee fields that the Arhuaco managed. The way I see it is that certifications exist today where relationships between grower and eater once did. Concerned consumers who don't know their farmers put their trust in a certifier who might put their trust in a third-party farm inspector. The more links in the chain between producer and consumer, the more likely trust can be broken, or one party can exploit another. But until and unless we are able to get to know the farmers who feed us, certifications, as problematic as they are, are our best tool to ensure our appetites don't further feed a system built on farmer exploitation.

A chicken ventured out of the shade of hazelnut branches bending with nuts to get some water. The sky was dark in the west and rumbling with thunder.

Regi's goals are ambitious. He had proven his system on this land and had already begun to build a larger Tree-Range farm on seventy-five acres, much of which was previously industrial corn and soybeans. Regi started by planting eighty-two hundred hazelnut trees.

## Back to the Land

The week after my visit, Regi would welcome a group of city folks who would pick out their own chickens to take home for dinner, but their visit was about much more than that. The visitors were wage earners living lives similar to the immigrants who work in chicken CAFOs and factories who have likely never seen a chicken that seeks the shade of a tree. People, who whether they are from a concrete jungle or an actual jungle, could, with help, step away from their low-earning jobs and adopt his techniques. Regi would explain his method, talk to them, and help process the birds. And instead of feeling like part of a system, they could feel like part of a *place*, and part of a chicken-farming insurgency.

# Part III: Regenerate

# Chapter 9
# Downstream People

# Mississinewa River, Indiana

**Regenerative Agriculture Is Knowing Your River**

I tripped over the corn but hung onto the surfboard.

"It's not a surfboard," Cliff said, from the nose. "It's a paddleboard."

I preferred my version. If Hawai'i had corn, couldn't Indiana have surfboards? The corn stalks and silage left from the previous harvest angled toward the sky like palm trees snapped in a hurricane.

My friend Cliff was indulging another one of my half-thought-out ideas. We walked out of my garage with a kayak and a "paddleboard," three days of provisions, and a goal of reaching a place that was sacred to those who knew this land best. Along the way I hoped to process my global regenerative journey, thinking about how it changed me and, after all this time, what the heck regenerative agriculture was and what it meant for the world.

In Colombia, the Arhuaco told me that every place has the magic. They said to tie the cotton bracelets they gave me onto trees when they came off. I had left one on a branch on Kaua'i and another on a sycamore my kids climb that leans over the river Cliff and I walked toward.

I had left Kaua'i asking myself a question that Malia Chun, a Native Hawaiian cultural practitioner, asks her students: What river do I belong to?

A large sycamore tree tilts across the Mississinewa River near where Kelsey and Cliff launched their adventure to get to know the river better. CLIFF RITCHEY

I suppose I belong to the Mississinewa River. If it would claim me or not—that's another question. Most of my life I didn't know it existed, despite being raised near its headwaters. When I peed off the front porch of my clubhouse in the middle of my grandpa's cornfield, it ran to the Mississinewa. When I bled from a fall, a baseball to the face, a go-kart crash, or a clod of mud thrown by my brother, my blood ran to the Mississinewa.

In high school, we moved to a different watershed, but the school was the same—Mississinawa Valley High School. I don't think I ever pondered what river had carved the valley the school was named after. In my defense, there really wasn't a valley at all. On most farms in the school district, farmers could leave their tractors in neutral and not worry about them rolling down a hill because there were no hills.

Note the "-awa" ending of the school name as opposed to the "-ewa." It might be a misspelling. I'm not sure. In any case, the word Mississinewa was made up by White folks. It's an adaptation of what the Myaamia (Miami People) called the river—Nimacihsinwi Siipiiwi.

When my wife and kids and I moved to our current home thirty miles downstream from where I grew up, I still didn't think much about the river, which was only a short walk away. There was a lot I didn't know about this land. In the beginning I didn't know where the deer like to bed at night. I didn't know that a beaver dam, which was slowly eroding away, had created the pond. I didn't know the name of the giant tree that looked out over it all. I didn't know the songs of different types of birds, where the groundhog lived, the areas that frosted first, what plants were poisonous to sheep, what birds eat chickens, or that black rat snakes climbed trees.

I loved the outdoors. But whenever I said that, I was thinking about oceans, beaches, and mountains. Indiana was a place you left to seek the Great Outdoors. Most definitely Indiana did not have the magic. But I didn't yet know this place we bought. It was ours. The property of Annie and Kelsey Timmerman. Twenty acres of too-swampy, too-hilly land that, in the words of Mark Shepard, had been "fucked over for two hundred years." I knew that when people asked how many acres we lived on, and when I told them twenty, I felt some sort of pride of owning so much. My tree. My lawn. My heron. My pond.

But this five-year regenerative journey changed all that. Now I feel something else. I think Malia would call it *'āina*, or love for the land, which requires *kuleana*, or responsibility and privilege

Kelsey and Cliff carry a paddleboard through a field to the Mississinewa River. In 1934 a historian wrote that "no other stream in Indiana, and possibly none in the rest of America, offers a more unique combination of scenic beauty and history interest." JULIE RITCHEY

to care for it. I've stepped into a relationship with this place I call home in a few unexpected ways.

We got sheep. We named them after our kids' great-grandmas—Betty, Clara, Rose, and Frances. I like to think I use holistic management as I rotationally graze them around our front yard. When I sent pictures of the sheep to Will Harris and to Dalmas, they both said nice things about my grass. I'm not sure they could've given me a higher compliment.

Early on in my shepherding days I was telling Will about the trouble I had moving the sheep: They were so athletic, so hard to catch, that I had to tackle them to get them to do what I wanted. One day I was chasing Clara and she nearly kicked Annie in the face. Will looked at me like I was a two-headed Yankee and explained that my sheep thought I was a predator because I was chasing them like I *was* a predator.

I've since improved my technique. Now, when we move the sheep, my daughter, Harper, shakes a Ball jar full of dried corn and they follow her—chasing or tackling not required. Recently, during the last days of fall, we were moving them to their winter home in the woods. The sun was setting, and the sheep were happy. I whistled, warbling like the Maasai, to them. Harper ran with her jar and the sheep skipped behind her. She said it was one of her favorite moments of the whole year.

The sheep have taught me many things in addition to humility. I now know all about the various grasses, wildflowers, and other plant life that grow in our yard, and I can talk for hours about how difficult it is to have and care for animals. Meanwhile, friends bring their children to meet them. And the sheep are keeping honeysuckle and autumn olive from taking over the

# Downstream People

pasture while I let black walnut trees climb toward the overstory of what I hope will become a multi-layered food forest.

We also added chickens. I'm the chief chicken tender, but Annie and the kids help, too. My son, Griffin, squeals and giggles when the rooster chases his sister. The chickens roam the woods, pasture, and edge of the pond during the day before retreating to their coop at night. One evening I was a little late putting them up, and they were nowhere to be seen. I figured something had gotten them, but as I walked around the pasture, I heard a tiny coo from overhead. Three of the chickens were perched in a young walnut tree. I thought of Regi's Tree-Range farms and how they are originally from the jungle and prefer the woods. The chickens have kept me in tune with the timing of sunsets. The raccoons and hawks have taught me how to think like a predator to protect the flock. The avian flu taught me how to administer mercy and end the suffering of sick birds.

We also added a compost pile and an expanded no-till garden, which has living roots in it year-round—strawberries, lettuce, peppers, carrots, potatoes, tomatoes, and plenty of volunteer weeds during the season and a cover crop of rye over the winter months. It's much more chaotic and less productive than Howard's garden in North Carolina, but we find joy in sharing the harvest. This system produces more than food; it strengthens relationships.

We had bees for a year. They were a friend's 4-H project. One day late in the summer I was on the back porch and heard a strange hum. I realized it was coming from the empty pond, blanketed with jewelweed and covered in bees stocking up for winter on one of summer's last flowers. It was as if the world

Kelsey's daughter, Harper, introduces the sheep to a new chicken. The sheep rotate around their pasture and the family's front yard.
KELSEY TIMMERMAN

## Downstream People

hummed, and I suppose it does, always, and I feel like I'm starting to hear it more.

The garden and the sheep both exist where lawn once did. Lawns, as mentioned earlier, are the number-one irrigated "crop" in the United States. A common criticism of regenerative agriculture is that its practices, such as rotational grazing, require more land to produce the same amount of food, yet we waste so much space, time, and energy maintaining lawns. I've committed to less mowing and more growing.

The more time I've spent on the land seeing and listening, the less the land feels like something that could be owned. And the more time I spend walking and learning from the land or sitting next to or floating down the Mississinewa River, the more my love for it has grown.

•

Cliff and I lowered the board on the edge of the river. I squinted to find the cotton bracelet in the sycamore tree, but it was gone. Maybe it let go and floated down the river. Maybe it was woven into a bird's nest.

We couldn't believe how much the river had changed. The day before, when we'd checked, the water was clear and calm. But now it was muddy and spiraled with eddies and frothed with foam. It seemed like a change we shouldn't have been surprised by. Like a thing a person should know was possible.

Rivers rise and fall, and we don't even think about it, but the people who have lived the longest on and near this place must've known. Must know. I hadn't thought enough about who had called this land home before me, but spending time with the Arhuaco, Tupinambá, Maasai, and Native Hawaiians led me to learn more about those who are indigenous to my own homeland.

The Delaware and Miami tribes called the banks of the Mississinewa home. Women planted and tended food forests of peaches, grapes, apples, persimmons. They cultivated fields of corn and wheat that stretched for miles along the river. They tended pigs and chickens. Men hunted and trapped. But also men would farm and women would hunt. Together they planned peace, governed their villages, looked after the sick, and told stories that connected them to the world.

•

The kayak was packed full, and I could barely get my legs in the cockpit. Cliff gave my stern a push and the nose immediately pointed downriver. I grabbed onto a fallen tree and waited for him. We had hoped for enough water to paddle, but this was more than we'd bargained for.

Cliff is a planner, which I'd relied upon many times during our travels, but I had handled the preparations for this trip. He asked questions I couldn't answer. What would the higher water level mean for our trip? Where would we camp? We had met a few times to go over satellite images and a random old blog post that covered a few sections of the river. But in many ways, it felt like we were making a first descent.

Cliff shoved off on the paddleboard and immediately his fin snagged on a tree beneath the surface.

I laughed. Here we were heading out on this big adventure and we were stuck before we even began. Cliff, who had much

## Downstream People

more experience on rivers, was less than amused, maybe even nervous. He was also more likely to believe in signs, and getting snagged this early did not bode well for our success.

I knew enough that if Cliff was worried, I should be too. When no one knows the river, no one knows when it's safe or dangerous. He got unsnagged and we both pushed into the current.

The river pumped past my neighbors' fields, splitting around islands, rapidly bending blind corners. We didn't talk much except to point out possible places we could get hung up or pushed against fallen trees, which could lead from tipping to pinning to death. Rivers bring life, and rivers take it away.

From the kayak I could only see slopes covered in cottonwoods and giant sycamores. A deer swam across and walked up the bank and gaped at us. I don't think she'd ever seen anyone paddleboard on a river either.

I had traced the Mississinewa enough on satellite images that I knew it was a tree-lined snake slithering through squared fields. Everything in the Midwest has right angles. The fields. The yards. The roads, most of them running north and south or east and west. Straight lines make defining ownership easier. Each square or rectangle on a plat book has a name written in the middle. Obviously, the Mississinewa, which you've likely never heard of, is not as mighty as the Amazon, but I couldn't help but think of Selma and Raquel in Brazil who fought against this Midwesternization of their forested homeland.

We crossed under bridges I had driven on countless times. It was another world below. The water had calmed and the trickling over small rapids and through fallen trees sounded like a relaxation track. A heron squawked and we followed it downstream until it got tired of the chase and circled around behind us and landed upriver. Then it was replaced by a bald eagle. We stopped paddling and let the current silently take us beneath its perch. The bird was about three feet tall, watching us like an armless toddler. It seemed big until it dropped from the tree and expanded its six-foot wings, and then it seemed mythical. The first two days on the river, we saw more eagles than people.

We crossed beneath the Cumberland Covered Bridge. This bridge near the small town of Matthews was such a big deal back when it was built in 1877 that it is still a big deal today. There's a park beside it where there's a festival each year in the bridge's honor. Kids can bounce on inflatables as adults eat corn dogs while walking through the exhibit of old tractors, even those powered by steam engine. Old tractors are a thing at county fairs, a connection to the past, antiquated centerpieces for conversations of how the world once was.

Other than the bridge, the most remarkable feature of the town is Matthews Feed and Grain. The company's silos and mill sit on the unusually wide main street. The mill was started as a co-op, meaning the local farmers owned and controlled it, making decisions by vote. Before rail came in, when grain was still transported by river, its shores were dotted with other small towns and their own mills, which have since disappeared.

As acreages and yields grew and family farms consolidated, the mill became a corporation. Today, it produces feed for CAFOs and show hogs—the swine shown at fairs that can sell for thousands of dollars. Matthews has a population of one hundred, twenty fewer than it did a century ago.

After spending a night on a sandbar, we reached the city of Marion, with a population 280 times the size. The Mississinewa straightened and widened, leaving no protection from headwinds, which were tougher on Cliff standing on the paddleboard. The wind carried the scent of dirt and oil. Signs on the banks cautioned against swimming or even wading in water for several days after a rain because of the sewage and wastewater runoff.

We paddled through the degeneration.

Once the smell of Marion was downwind and the river narrowed again, Cliff and I relaxed, chatting about our first jobs. I started lugging 2x4s and swinging a hammer for my parents' wood truss plant when I was fourteen. Getting up at 5:42 am and working in a world of splinters and sawdust was good for me, as suffering

Cliff likely holds the record for longest paddleboard journey on the Mississinewa River, fifty-plus miles to the Mississinewa Reservoir, which was created by a large dam built in 1967. KELSEY TIMMERMAN

often is. But strapping on a backpack and spending time in the woods and on rivers had been just as important to Cliff and me. The monotony of hiking built just as much character as the monotony of stacking lumber.

"So," I asked to the river as much as Cliff, "why is it that society doesn't value the type of thing we're doing right now in the same way as a job?"

"Money," Cliff said, his bushy hair, reminiscent of the Arhuacos', sticking out from beneath a faded trucker's hat. "This is a different type of work. It's not a work that rewards you financially, so therefore society doesn't value it."

In contrast, at a certain age the Miami sent young boys into the woods to pray and fast, sometimes for as long as six months. They painted their faces black with charcoal and waited for dreams that would guide their lives. They were sent out with no other humans, but they weren't alone. The Miami saw the plants and animals that surrounded them as their kin. They had a reciprocal and respectful relationship with nature, and through it, the Miami received food.

The Miami were in relationship with at least 161 different plants. They passed on these relationships from generation to generation. They knew it was time to hunt morels when lilac buds, mayapples, and plum tree blossoms appeared. It was time to plant corn when they heard the first call of the whippoorwill.

Corn has played an especially important role in Miami life for two thousand years or more. That's two thousand seasons of observation and experience. Roundup Ready corn, introduced in 1997, which surrounds the Mississinewa region today, has been planted for barely a quarter of a century, less than 5 percent of the time the Miami grew corn on these lands.

The Miami had at least seventy-one different terms related to corn. Growing it was about way more than food production. Harvesting, storing, and cooking corn represented an important link with the land and was critical to Miami identity. And they produced yields up to fifteen times higher than those of the European settlers who forced them out.

Like many of the other Indigenous people I had learned about on this journey, the Miami were regenerative by culture and necessity. They practiced continual cropping, meaning there were always living roots in the ground, so their crops outcompeted other plants. They didn't turn the soil over. All of this contributed to maintaining and building the organic matter of the soil, and likely fit even the strictest definitions of regenerative agriculture today. They belonged on these lands. However, there were other forces at work that would forcibly remove them.

In 1795, the United States government and an Indian confederation signed the Treaty of Greenville, in which much of Ohio and part of Indiana was ceded to the United States. Up to that point, the Native Americans had successfully protected their homelands. They continued to live there for a while, on land that was, in the eyes of the federal government, no longer theirs.

By 1812, it was clear that settlers wanted them gone. Recruitment campaigns called White men to arms with fear tactics: "Your wives and children, all you hold dear are in danger of being destroyed by savage violence." And future president William Henry Harrison, the former governor of the Indiana Territory, viewed the villages along the Mississinewa as "decidedly hostile" and said that the "whole of the provisions [of the villages] must therefore be destroyed and the houses burned."

On the night of December 17, 1812, while many Native men would've been out hunting, the Americans launched an attack. Seven men were killed and forty-two others, mainly women, children, and men too old to hunt, were captured. The soldiers burned their homes and destroyed their corn reserves. One of the American officers was disturbed by the conduct of the soldiers and said, "We shall suffer for this, we have not seen the end."

In retaliation, the Miami and the Delaware rounded up men and boys from two nearby communities. Outnumbered perhaps by a factor of ten, they fought so fiercely that the Americans retreated after an hour, taking the hostages with them, and claiming the destruction of the villages and the Battle of the Mississinewa as great victories.

Each fall there is a reenactment of the events along the Mississinewa, billed by the group Mississinewa 1812 as "America's Most Exciting Living History Weekend." Its website also claims the battle was a great victory and mentions how well the hostages who'd been separated from their families, their lands, their people, and their homes had been treated.

Indiana became a state in 1816. In his 1934 recounting of the aftermath of the Battle of the Mississinewa, historian Ross F. Lockridge wrote that "the Indian paradise on the Mississinewa was one of the last of the wilderness fastnesses of the middle west to be invaded by White civilization."

•

## Downstream People

"We should just camp here," Cliff said as we perched on a sliver of mud where the river widened into the reservoir.

We were now on federal land. The Mississinewa River was dammed in 1967 by the Army Corps of Engineers, creating the reservoir, Mississinewa Lake.

We had paddled more than twenty miles on our second day. All the beaches I had identified from satellite images as possible campsites had turned out to be covered in poison ivy or were actually cliffs that looked like beaches only from space. I cursed the limitations of the satellite information and my reliance upon it.

I felt bad. Cliff had been standing on the paddleboard all day; he was exhausted. And it was getting late. He still suffered some long-term effects of having COVID six months before. We were both hungry. It was one of those character-building, friendship-testing moments. I didn't know what to say. As he bit into a granola bar, I sank into the island. First, black water oozed between my toes, and then I couldn't see my toes. Sand flies circled my ankles like vultures.

I had run out of times that I could say, "A few miles more.... I think there is a spot." I had passed up a good one five miles earlier because it wasn't as good as the one we found on the first night. I was greedy and the river had changed.

But there was no way I was going to sleep on this mud bar and get eaten by flies all night.

"Okay," I said, spotting a boat ramp. "Let's stay right there." The Department of Natural Resources wasn't going to like it, but I agreed to pay the fine if we got caught.

He agreed. The only flat spot was in the turnaround area cut into the side of the cliff for trucks to back their boat trailers down the ramp. If anyone came for some night fishing, our tent would be in the way.

I hiked up the steep boat ramp away from the reservoir to get a phone signal to send updates to our families. At the top, the ground leveled off into a much better area to pitch a tent, but we would've been more easily discovered. It's then I realized where I stood. Up here the open fields were all managed by Mississinewa 1812. Once a year, tourists and vendors flooded to this spot to see actors dressed as American soldiers and Delaware and Miami Indians.

Kelsey starts dinner at their campsite on a boat ramp near the grounds of Mississinewa 1812, which features an annual reenactment of the battle between the US government and the tribes that lived along the river. CLIFF RITCHEY

**Downstream People**

You could nosh on a giant turkey leg or a funnel cake while touring the military camps or the "Native American Village" to experience what life would've been like along the river. However, much of where that living took place is underwater today because of the eight-thousand-foot-long earthen dam that exists in part because of what happened in 1812. The moments re-created here once a year represented a drastic, permanent change for people and the landscape of the Mississinewa River.

Lockridge wrote as such in his 1934 account:

> [The Mississinewa] is a scenic river. Nearly every mile of its course is made beautiful by winding curves, rippling rapids, craggy banks or short and pleasing straightaways, flanked by level and fertile bottomlands. No other stream in Indiana, and possibly none in the rest of America, offers a more unique combination of scenic beauty and history interest. It is a vital unchanged monument of a great and simple people. Along its shores a strong unit of a mighty race saw the summit of their power and there came their final passing as a race, all within the course of three-quarters of a century.

I appreciate the sentiment, but the more I've learned about the Miami and the other peoples who lived along the Mississinewa, the more I've come to realize he was wrong about two things. One, they weren't simple—but perhaps the settlers', the historians', and even our own understandings of them, were. And two, they haven't vanished.

Before the river trip I'd met with Daryl Baldwin, a linguist and member of the Miami Tribe of Oklahoma. Growing up, Daryl had ties to his Miami heritage, but not much knowledge of the culture. Few did. His generation looked like it would be the first to never hear the language of the Miami. Daryl said the language was sleeping. He played a critical part in waking it up.

"I wanted to fill a gap in my identity," he told me. "Our history had been erased."

By poring through archives, Daryl and other linguists were able to find the Myaamia language. Eventually, he started speaking the language himself and taught it to his kids.

Miami tribal leaders recognized the importance of Daryl's work and partnered with Miami University to create the Myaamia Center, which promotes the Myaamia language, culture, knowledge, and values. Daryl is the director.

Archived-based language revitalization is a new field. For his work, Daryl was named a MacArthur Fellow and awarded a "Genius Grant." Daryl and the Myaamia Center have shared their methods with 137 other tribes in an effort to return other languages from the archives to the people.

In the '90s Daryl led the first language camps, teaching the young and old how to speak their reawakening language, on the banks of the Mississinewa River at a sacred site known as the Seven Pillars.

•

After Cliff and I made our way across the reservoir and portaged around the dam back to the river, we eventually paddled to the Seven Pillars as well. It was our final destination.

The pillars—twenty-five feet tall and so wide that Cliff and I together couldn't reach around them—were cut into the

**Downstream People**

limestone cliff that had been deposited beneath a shallow ocean 425 million years ago. Each is shaped like an hourglass—the tops, even after hundreds of millions of years, still filled with more time than the bottoms. Instead of echoing the splashes of the small rapids we crossed, the alcoves behind the structures seemed to swallow sound. Three of the seven were next to each other like a grand porch, and the rest were a few strokes downriver.

The Miami started gathering at the Seven Pillars long before European contact. Where, according to writings from the Myaamia Center, they lit fires in the caves and told stories "of long-ago events ... where people sometimes interacted with the other-than-human beings who also were said to dwell there." As with many Indigenous cultures, the Miami's stories connect people with their ancestors and with the land itself. There's a season for storytelling, starting when the insects stop singing in the winter and ending when the frogs start singing in the spring.

They share a story of their origin. A telling delivered by Jerrid Baldwin, one of Daryl's grown children, begins like this: "At first the Miami came out of the water. And the place they came out of is called the 'coming out place.' 'Grab hold of tree limbs!' they yelled at each other. So they grabbed hold of those tree limbs. They pulled themselves out of the water. And right there along the bank of the river, they built a town."

Their origin began with nature saving them. Millennia after that and thirty-four years after the Battle of Mississinewa, on October 6, 1846, four miles downriver of the Seven Pillars, two hundred Miami were loaded onto canal boats on the Wabash River. They had been given a few minutes to round up their

The Seven Pillars, a sacred site to the Miami, who gathered and shared stories in caves behind these limestone formations long before European contact. CLIFF RITCHEY

possessions. A newspaper reporter recalled the scene: "I remember the sober, saddened faces, the profusion of tears, as I saw them hug to their bosoms a little handful of earth which they had gathered from the graves of their dead kindred. But stern fate made them succumb; and, as the canal boat that bore them to the Ohio River loosed her moorings, many a bystander was moved to tears at the evidences of grief he saw before him."

The journey by canal boat to Ohio, and then by riverboat to Saint Louis, and ultimately on foot to a reservation in Oklahoma, would take more than a month. At times the Miami people were listed not with the passengers on the manifest, but with the livestock and supplies. They were cargo. They were close to nature, and since nature was viewed by Americans as a thing that could be exploited and extracted, so could the Miami. So could the forests. So could the rivers. So could the soil. They were removed from this land and replaced by farmers, by people like me. People who asked the land to give and never gave back.

•

In my low moments, I fixate on how we are destroying the earth beneath our own feet. Selma and the Tupinambá face a monster in Brazil. Celestine and Dalmas face it in Kenya. It's in the dusty, lifeless deserts of off-season grain fields around my house, in the air and water my family relies on. It's nearer and more of a threat to some of us than others, but it surrounds us all.

Fossil fuels and the machines they churn have disconnected our feet from the soil. Artificial fertilizers have disconnected animals from the farm. Pesticides have disconnected us from recognizing— and even having—healthy ecosystems that teem with a balance of

## Downstream People

insects, microorganisms, and other life. Overreliance on experts with one-size-fits-all technologies has disconnected us from the knowledge of our neighbors and our ancestors. In all of the places where relationships once existed, products now do. While corporations have profited, we've lost our way.

Will we find it? I don't know. It's so damn complicated. Each day I suffer cognitive dissonance as I simultaneously hold the awareness of the destruction we face and the renewal that the people I've met in my research work toward.

Sometimes I wonder what our purpose is on this planet, anyway. We don't photosynthesize, we don't pollinate, we don't fix nitrogen, we don't produce oxygen or sequester carbon. Yet the people I met wake up each morning, and they put their hands in the soil, their feet on the ground, and they get to work. Their work and lives prove that humans can be a force for good.

Recently, I heard CMarie Fuhrman, an Indigenous poet, speak to this perspective—and give me hope. She said that humans actually have a very important role to play, one that Indigenous peoples have served for thousands of years: Our purpose is to love and wonder and celebrate Earth and the life on it.

To the Arhuaco, the place they live is the heart of the world. What if we all treated our homelands with as much respect and care? What if our active participation, love, and interaction with our places allowed them and us to thrive? It's human involvement with plants and animals that builds soil on Will's land, that fights desertification in Kenya, that leaves healthy land for the next generation of farmers like Luke and the Wepkings, that restarts rivers in the Amazon and on the Maasai Mara. Maybe the heart of the world isn't just a place. Maybe it's a living thing.

Maybe it's the Arhuaco. Maybe it's all of us. Maybe we are the heart of the Earth.

What if we acted as if we were that important, our actions that consequential, and had the audacity to pursue a life's work that can't be accomplished in our lifetime, and planted something that would outlive us? What if we, like so many of the farmers I met on this journey, waged small acts of hope that reject the idea that we're doomed?

What if we actually paid attention?

All this hubbub about regenerative agriculture. Maybe we're laying too much on the laps of farmers. We need to think beyond farming regeneratively to *living* regeneratively, whether or not we are farmers. That can start by, as Regi put it, decolonizing our minds and seeing ourselves as part of nature, belonging to it, not above it, not dominating. We could get to know a neighbor, a river, and learn from the Indigenous communities who were here before we arrived.

The global community needs us all to fight for our local air, water, plants, animals, and people.

The regenerative movement is growing in popularity. I see it mentioned in many more documentaries, news stories, and books. The word "regenerative" is showing up on products in grocery stores, even in Indiana. Conventional farmers who have experienced more weather disasters, rising costs of fertilizers and 'cides, and continually unreliable prices for their crops are beginning to question the industrial system. Some farms are making changes. Between 2012 and 2017, the number of acres using cover crops doubled. No-till is now used on a third of cropland acres in the United States, although often supported by healthy doses of 'cides.

However, some of the original leaders of the organic farming movement are concerned about the misuse of the term "regenerative agriculture"—it's often undermined by the largest proponents of the industrial model by referring to the adoption of a single practice, such as cover cropping or no-till. Regi and Will Harris also spoke to me of this danger, which is real.

It all reminds me of what farmer and author Wendell Berry had to say about movements in general: "The worst danger may be that a movement will lose its language either to its own confusion about meaning and practice, or to preemption by its enemies.... Once we allow our language to mean anything that anybody wants it to mean, it becomes impossible to mean what we say."

"Regenerative agriculture" as a term has to mean something, or it will mean nothing. At the store, there are labels and certifications that can help us out, such as the Real Organic label and Regenerative Organic Certification. Good standards can meet Will Harris's definition that we need to restart the cycles of nature.

But regenerative practices and outcomes look different in different ecosystems, cultures, countries, and communities. That's why, after thousands of miles and zillions of hours on my journey to define the term, I've given up. I've decided that it's impossible to define regenerative agriculture. Still, again, as my friend Michael O'Donnell told me, "I know when I see it. When I feel it."

With that in mind, I present this list of certain ideals I've seen and felt that have helped me know when a farmer is practicing agriculture that is actually regenerative. These farmers:

1. **promote diversity** by growing a variety of crops and/or animals that support and are supported by thriving communities of insects, birds, microorganisms, and other plant and animal life.

2. acknowledge that **soil is alive** and healthier when it's rich with plants and living roots.

3. **try not to disturb the soil by plow or chemical** to avoid breaking its web of connections.

4. **work with animals**, wild and domesticated, in ways that give back to the web of life.

5. **reject all or some of industrial agriculture** and the consolidation, centralization, commodification, and disconnection it demands.

6. **embrace a connection with nature** based on gratitude, reciprocity, responsibility, and often spirituality.

7. believe that **humanity can be a force for good**, renewing and regenerating the land, our spirits, and our communities, if we treat each other and our natural ecosystems with respect.

●

I hold onto hope that we can change. That hope is buoyed by those I've met on this journey, people who are taking on the powerful forces of commodification, consolidation, industrialization, greed, and are rebuilding lost connections.

## Downstream People

This is global work that requires us each to be regenerative champions in our local communities.

Sometimes that can look like not building a concrete dam and trying to work with beaver instead. When the beaver dam that was here when we bought the land washed out, our pond disappeared. A USDA agent who visited suggested that if I wanted it to return, I could pour a concrete spillway. Made from rock and sand harvested from who knows where, it would cost more than $50,000.

The USDA would contribute a few thousand dollars because slowing down the journey of the pond water, which was fed by 750 acres of polluted farm runoff, to the Mississinewa River was healthy for the environment.

"A beaver dam would be free," I had told him.

Even if I could've afforded a concrete dam, I wouldn't have built one. Instead, there I was, pounding large wooden fence posts into the muck of the creek bed until the tops splintered beneath the onslaught from my mini-sledgehammer. My forearms burned. My rubber boots had long since filled with water and I'd opted to go barefoot. As I gradually sank deeper, reaching the top of the post became more exhausting. When the post wouldn't go any further, I moved on to another.

I wiped sweat from my eyes with a forearm spritzed with muck. A heron flew over, either trying to inspire courage or simply checking to see if I was done, so that it could return. A muskrat had built a den in the middle of what was left of the pond and the heron had taken a liking to perching on it.

My friend Scott—who was among the few people who appreciated what I was trying to do instead of reminding me that most people are trying to get rid of beavers or that it's illegal to transport beavers across county lines—tossed down bundles of willows he'd trimmed from the banks, and I threaded them through the posts. After hours of work, a woven wall of willow separated the pond from the creek.

This was beaver work, and a beaver would be much better at it. The beaver dam that had once stood here had been twice as tall, and a grown man could walk across it as if it were solid ground. But the previous homeowners, concerned about flooding, had killed the beaver that had built it.

Turns out beaver dams do a lot more than slow the journey of polluted water. The ponds that beavers create encourage water to sink into the ground and replenish aquifers. By collecting water and releasing it gradually, they dissipate flooding during extreme weather. Insects live in their dams, and where there are beavers, there are more grassy areas for duck nests, willows for songbirds, more turtles, lizards, and fish. Beaver ponds prevent soil erosion, capture carbon, and create wetlands that act as barriers to wildfires.

My willow wall is known in the "Beaver Believer" community as a "beaver dam analogue." The hope is that the wall will raise the water level enough that a beaver could see the potential for a den here, notice the abundance of willow and poplar on the banks of the pond, and take over the dam-building process where Scott and I left off.

My willow wall made the water rise. But then it washed out, pushing the posts over. I rebuilt it. It washed out again.

My efforts remind me of a story that many Native American tribes share in various versions. Such stories are shaped by the teller and the audience. A book isn't quite as alive as an oral story, so I'll do my best to leave you plenty of room to co-create. It's called "The Old Woman Who Weaves the World" and goes something like this:

> There's an old woman in a cave along the river not too far from your home. She has hands like your grandma's. She's been weaving for a long time, for as long as the river has run. She's almost done with her creation. It's the most beautiful and complex garment the world has ever seen. She's contemplating how to finish it. How to make it perfect. She's also cooking a stew deeper in the cave, because even old women who weave the world multi-task. The stew needs stirring, so she leaves the garment on her seat to deal with it. The stew is in a giant cauldron made from the mud of the river and contains seeds from every plant on Earth. If she goes too long without stirring it, the seeds could get scorched, and then who knows what trouble might befall the world. So, she weaves and stirs. Stirs and weaves. She shares the cave with a dog. The dog sort of looks like your favorite dog when it was a puppy, but the dog is not a puppy. When the woman gets up to stir the soup, the dog perks up, walks over to her nearly finished, perfect garment, grabs a loose thread, and runs around the cave. When the woman returns, she stares at the chaotic mess of the unraveled garment on the floor. She stands quietly, pondering, before bending to grab a single thread. As she sits down and picks up her needle, she imagines a new design and gets back to work.

This is the work: people carrying out acts of regeneration in a degenerated world, recognizing the harm that's been done and, despite it, imagining and working toward a new world.

I'll rebuild the willow wall again. This time I think I'll add some clay and rock at the base. And the other day I discovered a new beaver den on the Mississinewa not far away. The row of chewed trees had begun to extend to our property. The beavers are returning in their own time, but I can help them along.

These small little steps of restarting the cycles of nature, as Will would put it, have helped me connect with where I live. I used to walk outside to be alone, to clear my head. That's not the case anymore. Now, when I walk outside, I see the abundance of life, the cycles of nature that it's a part of, and I recognize that I'm a part of as well. I check in on the heron, the progress of the new muskrat den, the robin's nest, the growing rabbit family in my failed pumpkin patch.

The cycles of nature are in us and we are in the cycles of nature. We are part of the water cycle and nutrient cycle. We breathe in oxygen produced by plants and exhale carbon dioxide that plants breathe. When you think about it, you can't plant and nurture a seed, or tend to a sheep's lame foot, or take a bite of food without realizing that you're dealing with every cycle of life and law of the universe.

Originally, I'd thought that regenerative agriculture was all about using particular agriculture practices to sequester carbon. But now I know that it isn't that simple. In the Anthropocene—the age in which humans have become a geologic force—it's about living with the changes, recognizing the challenges, yet still bending down to pick up the thread of hope, or the mini-sledgehammer, and

reimagining what the world could be. It's okay to mourn what has been lost, to fight like hell to keep more from being lost, just as long as we continue to get back to work rebuilding and regenerating.

A book won't change the world. Hope alone won't, either.

Action will.

Our hunger is a gift that connects us to chloroplasts, lions, mycorrhizal fungi, and farmers around the world. And gifts come with responsibilities.

Regeneration is about so much more than agriculture. It is a choice, a movement, a philosophy, a decision we must make together. So, eat, breathe, plant, nurture, and embrace the abundance and wonder of nature like the world depends on it....

Kelsey's best attempt at building a beaver dam. Not even close to beaver-quality work. The first big rain washed it out. Yet he's undeterred. Anyone know where a fella can get a beaver?
KELSEY TIMMERMAN

Because it does.

•

After taking in the Seven Pillars, Cliff and I pulled the kayak and paddleboard onto the opposite bank. His wife, Julie, took a few photos of us. Like the end of any adventure, there was a sense of accomplishment, but I wasn't sure how to feel.

I lay in the water, anchoring myself in the middle of the river with my hands on the silted and rocky bottom. I let the river current lift my body like air beneath a wing. We'd paddled more than fifty miles downstream from the cornfield behind my home.

I attended Mississinawa Valley High School. Greenville is the seat of the county where I grew up. I have two degrees from Miami University. The year before I started, the university's mascot was still the Redskins. And I live on the banks of the Mississinewa River where the Miami once lived.

I recognize the horrible events, the removal and genocide of the Miami, and feel a sadness. I am a descendent of those who perpetrated and benefited from the removal of the Miami and other tribes, and I feel guilt.

I wonder what it would be like to know that my ancestors thousands of years back sat in this same place and felt the joys and pains of life. I feel longing, unsettled, rootless. I can't be indigenous to my homeland.

My ancestors forced those who are truly indigenous from this place, but they are not gone. Through the work of people like Daryl Baldwin, the Miami—the Myaamia—are remembering, and connecting, with places, people, and stories.

## Downstream People

A loose translation of Myaamia is "downstream people."

The current of the Mississinewa buried my hands in sediment, submerging me further in the dirt and pollution of our extractive interactions. The Miami, the river, the eagles, the heron, all reminded me that a different relationship with the land was possible. Ignorance, greed, and a lack of humility have disconnected us from the land and each other, but the act of getting to know what river I belonged to and the history of the people who know this place best—and my entire regenerative journey—felt like an important first step, a remembering, a knowing, a recognition of and responsibility to the people who came before and those who will come after.

Upstream, there are those who made decisions and lived lives that impact ours today. Downstream, there are people who will feel the impacts of our decisions and lives. We are connected, our interests intertwined.

The Mississinewa runs to the Wabash, which runs to the Ohio, which runs to the Mississippi, which runs to the Gulf of Mexico, which runs to the ocean and to every river in the world.

# Other Farms

For the five years I've worked on this book, I've traveled around the world to learn from farmers. Although you've met many of them, quite a few others welcomed me onto their farms and into their lives and helped to shape how I feel and think about regenerative agriculture. But because of narrative limitations—and the fact that you probably wouldn't want to read or even hold a book twice this size (seriously, the first draft was double what you're holding now)—I've had to leave a few out.

Thankfully, I have a little space here to introduce you to some of these pioneering folks.

### Huerto Cuatro Estaciones (Puerto Guadal, Chile)

Four Seasons Farm (Huerto Cuatro Estaciones) is like Howard Allen's in North Carolina, inspired by uber-productive and successful market gardeners such as Eliot Coleman and Jean-Martin Fortier. It's a great example of how farmers who adopt regenerative practices can be profitable and feed their communities on smaller tracts of land even when faced with the weather extremes of a place like the southern tip of South America.

Friends Javier Soler and Francisco Vio were both educated in the ways of and employed by industrial agriculture. The more Javier learned about growing organically, the more concerned he became about the health impacts of chemical-intensive agriculture. As for Francisco, his father was a crop duster who survived a plane crash but not the exposure to what he'd been spraying.

Javier and Francisco first started farming together in Chile's Patagonia National Park where they tended a garden that fed tourists and park staff before they branched out on their own

PREVIOUS SPREAD: Doña Rosa's farm in Puerto Guadal, Chile, near Huerto Cuatro Estaciones, uses regenerative practices—and it is not for sale! Doña Rosa and her farming partner sell at the local farmers market. CLIFF RITCHEY

ABOVE: Francisco's wife, Maria Jesus May, feeds the chickens and tries to figure out which of their young hens is a mystery rooster. Since this photo was taken they've welcomed two children onto the farm. CLIFF RITCHEY

to a beautiful setting next to Lake General Carrera near Puerto Guadal in the Aysén region. Thirty-five different types of fruits and vegetables grow in just about one acre of beds.

Throughout the growing season, they welcome interns from around the world. I spent several days there, planting crooked rows of veggies, and learned how to prepare the soil with compost and a broad fork. I incorporated these practices in my garden back home.

Before the farm came to be in 2018, many people in the region had stopped growing their own food—or forgotten how to—and had become totally reliant on food delivered from who knows where via unreliable roads. What I loved most about Huerto Cuatro Estaciones is that it helped its neighbors reconnect with their land and farm roots.

At the local farmers market, the only other grower, Antoine, had been a student of Javier's and Francisco's. A former oil executive from France, he farmed with Doña Rosa, whose grandmother was a famous healer in the region.

"My generation lost the knowledge of other generations," Doña Rosa told me. She was offered a lot of money for her land, but she turned it down. She explained, "This land is a gift, and I don't want to sell it."

"She's rich and she knows it," Francisco told me.

To learn more and apply to volunteer at the most beautiful farm I've ever visited, go to huertocuatroestaciones.cl.

### Sequatchie Cove Farm (Sequatchie, Tennessee)

Four generations from age two to ninety-two live and work on this diverse three-hundred-acre farm outside Chattanooga, Tennessee.

## Other Farms

What started as Jim and Emily Wright's retirement project became a radical reimagining for them, their daughter, Miriam, and her family.

I visited one morning driving back to Indiana after visiting Will Harris. Miriam's husband, Bill, and son, Kelsey, had some great Will Harris stories. Everyone bucking the industrial system to work more closely with nature seems to know each other.

The farm is set back from a winding road on the Cumberland Plateau, surrounded by Tennessee wilderness enjoyed by climbers, mountain bikers, and hikers. I had found them on a map of regenerative farms maintained by the Organic Consumers Association (https://organicconsumers.org/regenerative-farm-map/). Kelsey and I bonded over the fact we're both dudes named Kelsey, and he invited me to visit. I arrived as a haunting fog lifted to reveal pastures of sheep and several homes spaced out across the surrounding hillside. Miriam was foraging for mushrooms in the forest. Kelsey's kids and wife were in their large garden, gathering organic vegetables for their table, as well as for several local restaurants and markets.

Four generations worked together. "A farm like this needs young people and elders," Bill said, acknowledging the importance of having vitality alongside wisdom.

Kelsey and Bill are always learning, adjusting, and trying out new things, such as the field of small grains they hoped to harvest with an ancient combine. They rotate sheep, chickens, and cattle. Kelsey showed me one of his favorite spots—a swimming hole in the Little Sequatchie River. The river emerges from mountain springs and is the best place to cool off on hot days.

Jim, a former architect, was living with Alzheimer's and could no longer fully participate on the farm. They still include him, however. He helps snap green beans and sort through what his wife forages.

The farm was a mash-up of the Wepkings', Mark Shepard's, and Will Harris's in what they grew, how they grew it, and how a visit made me feel: I left questioning our industrial food system, and the fact we often forget about addressing the generational aspects of regeneration. Great-grandkids working and playing alongside their great-grandparents made me reevaluate our culture of single-generational living and how we care for the elderly.

Kelsey said that farming how they do isn't easy, and if things don't go well, he might have to get a different job. That hasn't happened yet, and he's eager to help others get started. "There's a lot of opportunity," Kelsey said. "Just this week I've had another new farmer raise a batch of chickens for us. I think, other than rich people farming for the fun of it, the only real future for farms of this size is working together."

Sequatchie Cove, featured in the documentary *Roots So Deep (you can see the devil down there)*, regularly hosts meals and workshops.

To learn more, visit sequatchiecovefarm.com and watch *Roots So Deep* at rootssodeep.org.

### Markegard Family Grass-Fed (Half Moon Bay, California)

Sometimes when rancher Doniga Markegard looks out from the headlands where her herds graze, she sees whales breaching in the Pacific Ocean.

**Other Farms**

As a girl, Doniga attended a wilderness school, where she learned how to track animals. Later she studied permaculture. She hadn't ever thought of ranchers as environmentalists, but then she met her husband, Erik, who showed her a thriving ecosystem where cattle mimicked the large herds of antelope and elk that once roamed the California grasslands.

Erik managed Neil Young's ranch. (Yes, the legendary rocker. Doniga and Erik toured with Neil through Europe acting out the stories in his songs.)

Eventually the couple found their way to their own 1,000-acre ranch nearby, south of San Francisco. They manage grass-fed beef and lamb, as well as pasture-raised chicken and pork, and they sell through their CSA and Bay Area farmers markets. Theirs is the first ranch in California to earn the Audubon "Bird-friendly Land" certification through the organization's Conservation Ranching Initiative. In total, the Markegards manage 11,000 acres along the coast.

My daughter, Harper, and I "helped" Doniga, her daughters, and their ranch hand Sue herd cattle, collect eggs, and move the chickens. Doniga taught me how to open a bag of feed (easily pulling out the stitching that seals it shut as opposed to my strategy of pulling and cussing to no avail before finally walking to the house to grab a knife) and about animal flight zones.

As a teenager, Doniga was mentored by Lakota elder Gilbert Tatanka Mani, who taught her the Seven Sacred Rites of the Lakota, the Sioux spiritual observances that focus on inner peace, joy, compassion, and connecting spiritually with nature. When Harper and I joined the Markegards for dinner at their home, one of their daughters chose the rite to read that night.

Some who are concerned about the crises facing nature believe we should walk away, stop messing with it. Doniga may have been like that once, but now she sees we don't have time. The native animals that once managed the cycles of nature have been hunted and pushed from the land by humans. Now it's up to humans to fix it.

"We need to have an ecological perspective," she said. "If we walk away, nature will eventually balance itself out. However, that might be at the risk of the destruction of our own species."

Doniga writes about her life in the wilderness and on the ranch in her books *Dawn Again: Tracking the Wisdom of the Wild* and *Wolf Girl: Finding Myself in the Wild*. The Markegards were also prominently featured in the regenerative agriculture documentary *Kiss the Ground*.

At the Markegard Family farm you can herd cattle while watching a whale breach. Kelsey helped Doniga and ranch hand Sue with the cattle, and Harper got a taste of an idyllic childhood as she played with Doniga's daughters. KELSEY TIMMERMAN

## Other Farms

To learn more, visit markegardfamily.com, read *Dawn Again* and *Wolf Girl*, and watch *Kiss the Ground* (kissthegroundmovie.com).

### Radiance Dairy (Fairfield, Iowa)

Grass doesn't walk; cows do. Yet the industrial dairy model moves grass and keeps cows in one place. By contrast, that's not what happens on the 173-acre Radiance Dairy Farm operated by Susan and Francis Thicke.

Francis and I stood in a pasture in southern Iowa that was formerly corn and soybeans and watched his cows munch on thick grass. The cows paid for their future meals by depositing a little fertilizer. No need to waste fuel or time to mow and bale or buy fertilizer. The majority of the milk, cheese, and yogurt from Radiance Dairy is consumed within five miles of the farm.

Such an operation requires an unusual community to support it. And Fairfield's status as such was certified by none other than Oprah, who featured it on her network's show *America's Most Unusual Town*. In the 1970s a popular yogi and transcendental meditation (TM) teacher started Maharishi University in Fairfield. People came, and then they stayed, embedding themselves in the community, now almost 10,000 strong, starting businesses and homesteads. In other words, there are a lot of folks who value local, organic food that's raised in harmony with nature.

As Francis and I watched a calf suckle on its patient mother, he shared that he's been practicing TM for years. "I'm seventy years old and feel like I'm twenty-five," he told me.

When a cow is unwell, he and his wife treat it with Reiki, a Japanese energy-healing technique.

But he hasn't entirely given up on what he learned in his previous career, in which he worked as a soil scientist for the USDA. "I straddle both worlds," he told me, focusing his observation on his cattle and the land with some surprising results.

For example, Jersey bulls are often said to be aggressive. Hypothesizing that the behavior may be caused by an early separation of calves from mothers, he started keeping them together longer. Now his Jersey bulls are calm.

Francis was an early subscriber to Rodale's organic farming and gardening magazine and was involved with the organic movement. He even served on the National Organic Standards Board from 2010 to 2017 but, like many farmers, Francis became disillusioned with the USDA standards. At his last meeting, he gave a speech about industry's corruption of the standards and his support of the add-on certification offered by the Real Organic Project.

"For years we were always looking at how we need to regulate agriculture," he told me, "but we're not getting there. Now, I'm encouraged that there is enough payback just from farming in the right way that people are going to start doing it."

If agriculture is to be regenerative, it will rely on folks like Francis, who straddle science and spirituality, activism and policy, and who have the patience to be a part of long-term, multi-level change.

To learn more, visit Radiance Dairy on Facebook and read *A New Vision for Iowa Food and Agriculture* by Francis Thicke.

### Janie's Mill (Ashkum, Illinois)

I rode shotgun with Harold Wilken as he navigated the Chicago suburbs in a box truck loaded with stone-milled organic flour

and other grains. We visited bakers, restaurants, breweries, and distilleries. Harold is a grain farmer and miller south of Chicago, and he takes pride in his work. He joked, "If we can't feed you, we'll get you drunk!"

Harold started farming when he was a kid and loved everything about it: the smell of diesel fuel, the plow turning over soil, riding in the tractor, and listening to rock 'n' roll. He passed that love on to his kids, including his teenage daughter Janie. There wasn't anything on the farm she couldn't do, he bragged.

In 2001, Janie was riding down a rural gravel road in a friend's car when it spun out of control, and she was killed. Harold's world

A field of organic soybeans on Harold Wilken's farm south of Chicago. In 2017, Harold started Janie's Mill, named after his daughter, who died in a car crash. Now he supplies local bakers and brewers. "I use nature to do nature," Harold said. KELSEY TIMMERMAN

was turned upside down. He realized that conventional farming didn't feel right any longer. He felt disconnected from the land and consumers. He wanted to invest in people.

He was the first in his county to start farming without chemical inputs. He went from farming 3,000 acres conventionally with a little part-time help to farming three thousand acres with seven full-time employees.

"Now I use nature to do nature," Harold explained.

One day he was watching his grain fill a trailer destined for a distant locale and thought, "Really?! My grain has to go seven hundred miles to feed a chicken, when there's all these millions of people nearby who eat bread?" He connected with bakers and with other millers to learn how to mill. In 2017, Janie's Mill was born.

He named the business after his daughter because, since her passing, he's always felt like she was guiding his journey.

To be a miller is to always be slightly coated with flour. Same goes for being a baker. So everywhere we stopped on his delivery route, Harold was greeted with a handshake or a hug, after which a cloud of Janie's flour would hang in the air. It swirled around us. The scent of hand-kneaded bread filled us.

In order for a regenerative food movement to grow, we need folks like Harold and businesses like Janie's Mill to provide the processing and relationship-building to connect farmers who want to adhere to regenerative ideals to consumers who value them.

To learn more and to begin your bread-baking journey, visit janiesmill.com.

**Other Farms**

### Perennial Pantry/Sprowt Labs (Minneapolis, Minnesota)

This isn't a farm, but what's happening in this nondescript office park outside Minneapolis will, with any luck, impact the future of agriculture.

Christopher Abbott has the climate solution pulled up on his laptop. The data is from the Intergovernmental Panel on Climate Change, but I get the feeling that the graph is his own, not because it's amateurish—quite the opposite—but because the bars match the colors of his company's logo, and because Christopher is a nerd like that.

Christopher's graph shows a range of all options available to take carbon out of the atmosphere, from sci-fi–flavored high-tech (and high-price) possibilities to soil carbon sequestration, which offers the most potential and costs the least. Our climate crisis is at the stage where we shouldn't rule anything out, but, using plants to grab carbon out of the atmosphere and deposit it in the land makes a lot of sense to me.

Christopher believes his company plays an important part in the climate fight since it produces and sells crackers, flour, and pasta made from Kernza. The more demand for products from carbon-sequestering perennial agriculture, the more the potential of the soil to store carbon is realized.

Perennial Pantry started as a college business plan and went on to win a competition. The latest version starts with a video, which Christopher played on his laptop.

Cue ominous music, clips of industrial agriculture gone wild—such as sixteen combines harvesting a massive field in a diagonal line—and Video Christopher intoning that "our farming practices have eroded the wealth of our soils in a system that contributes a quarter of global carbon emissions. We need a new agriculture."

Cue his tone rising with hope, the music becoming more inspiring, and pictures of people on the land and on computers doing research.

A renewed Video Christopher continued, "We can fight climate change by returning lost carbon to our soils through perennial crops. Perennial Foods is on a mission to decarbonize agriculture and re-carbonize our soil. We're building the perennial agricultural system, the future of farming, by bringing delicious, climate-positive food products to market.... Now is a singular moment to reimagine agriculture. Let's keep farming forever."

People like Christopher and companies like Perennial Pantry and Sprowt Labs, which designs processing equipment especially for Kernza, are working to develop infrastructure and markets joining researchers such as those at The Land Institute with entrepreneurs, academics, and farmers like Carmen Fernholz, imagining and working toward perennial agriculture, and helping develop legislation that would support it.

Because of them, it's now possible to start your day with Kernza pancakes for breakfast and finish it with Kernza beer—and help sequester carbon at the same time.

To learn more, visit perennial-pantry.com.

### Jason Mauck (Gaston, Indiana)

Jason is my neighbor, but I didn't know about him until I saw him interviewed by Chris Matthews on MSNBC in 2019. That same

## Other Farms

night I sent him a message, and within the hour I was sitting next to him on his tractor planting beans at midnight and drinking the beers I'd brought.

Jason had been on TV talking about the 2019 tariffs' impact on corn and soybean prices, but the reason he was on the media's radar in the first place was because years before, he'd started to, as he puts it, "farm weird." While he's still stuck in the mainstream agricultural system, farming large monocrops of corn and soybeans, he also plants soybeans right into his winter wheat, a practice known as relay cropping. By early summer his fields are striped gold and green, the plants getting their photosynthesis on with fewer chemical inputs. He harvests the wheat while the soybeans are still low, and once they have more room to spread, they produce more beans per plant than your granddaddy's soybeans. His wheat grows like a well-tended chia pet. And his fields retain more water and topsoil, since they always have a crop in them. If Jason were to tell you about all this, he would litter his description with entertaining and sometimes head-scratching images, talking about kernels of corn having daddy issues, for example.

In this book I've said that industrial agriculture "farms farmers." That came from Jason. He's got a way with words.

In addition to his relay crops, he started a small organic farm he calls the "funny farm" where he grows popcorn and hemp, and raises pigs in the woods. The pigs cleared out invasive species before becoming bacon that Jason was proud to share with others. He's part of a team that designed a solar-powered automatic "chicken tractor," a movable coop.

It's not always been easy for him. There are family and social pressures—and hardly any locals attend his open houses, though farmers from across the United States and other countries show up to learn from him. There's also the economic reality that regenerative agriculture often doesn't produce the financial growth of industrial models.

But the more I've seen Jason work with the cycles of nature, the more I've seen him come alive. This gives me hope that any farmer, regardless of how deeply entrenched in industrial agriculture, who experiences the joys and challenges of working on the land can change. And I've also realized that it's important that we give farmers the encouragement, space, and time to do so.

Follow Jason on X at @jasonmauck1.

**Grain Place Foods (Marquette, Nebraska)**

The first regenerative organic-certified food I ever ate was popcorn grown and processed by David Vetter. He handed me the bag while we stood in the facility he and his father, Don, started in the 1980s. Back then, they struggled to find anyone to clean their organic grain. Today they buy, clean, and store grain from more than a hundred small-scale organic farms.

After Don returned from WWII, he adopted the industrial farming practices promoted by universities and the budding agrochemical companies. But two years later, he noticed the soil was less alive and there were fewer birds. He can actually remember exactly where he was, driving the tractor, when he decided to change course. This was in the 1950s, well before it was seen as fashionable or practical to reject the new miracles of conventional agriculture.

"I don't want to be a braggart and go to town and say I raised three hundred bushel corn at the expense of the land and future

## Other Farms

generations," Don, who passed away in 2015, said in *Dreaming of a Vetter World*, a documentary about his family's farm.

Their neighbors wondered about the Vetters' sanity. David had gone off to earn a degree in soil science and attend seminary—where he focused his ministry on the soil—before returning to the farm. Then, after a few successful seasons of organic growing, the local paper ran a story about him headlined "Successful Organic Farmer No Longer Crazy." Today, the Vetters farm 280 acres on a nine-year rotation incorporating organic heirloom barley, soybeans, popcorn, corn, and grass-finished beef.

The Vetters were farming organically and regeneratively for decades before the USDA Organic certification came about. Because of his expertise, David was included in early meetings with Senator Patrick Leahy to discuss the Organic Foods Production Act (OFPA) of 1990.

Given that none of the next generation are interested in boots-on-the-ground farming, the family created the Grain Place Foundation and granted it majority ownership of the farm and Grain Place Foods. The foundation seeks to pass the Vetters's intergenerational farming knowledge to volunteers, interns, and future farmers.

Learn more about the Vetters and Grain Place Foods by visiting grainplacefoods.com or watching *Dreaming of a Vetter World* (dreamingofavetterworld.com).

# Resources

Continue your own regenerative journey and scan the QR code below. It'll take you to a list of organizations doing good work, as well as a description of some of the many certifications out there. I'm also sharing some books, movies, and podcasts that I find inspiring and interesting, as well as a list of links to all the farms in the book.

In addition to traipsing around the world, I did a lot of research for this book, which you can follow by checking out my endnotes.

This page also has a map showing the location of all the farms I discuss, a reader's guide, the ebook, and the audiobook.

And finally, if you have any questions, comments, or surefire plans on how to build a beaver dam, please email me at kelseywtimmerman@gmail.com, and follow along at kelseyt.com.

# Acknowledgments

I couldn't have collected and shared the stories in this book without the generous help of the farmers who allowed me onto their lands and into their lives. Learning to see from their points of view not only shaped this project, but forever changed the way I see the world.

I'm especially grateful to the Indigenous people who shared their time, knowledge, and stories with me. Those individuals include, but are not limited to, Dodoringumu Miguel Chaparro I of the Arhuaco in Colombia; Kaina Makua, Peleke Flores, and Malia Chun of Hawai'i; Dr. Raquel Tupinambá of the Tupinambá in Brazil; and Daryl Baldwin and George Ironstrack of the Miami Tribe of Oklahoma. They helped shape the journeys I recount in this book and also gave feedback on early drafts. When an author writes about cultures other than their own, they bear special responsibilities, including ensuring cultural accuracy and cultural sensitivity. I've made every effort to meet these responsibilities, and it would've been impossible without their help.

The book also benefited from the feedback of my fellow creative writers at Miami University, with special thanks to Rachel Grimm, who spent time going through an early mega-messy draft. Miami professors Daisy Hernández, TaraShea Nesbit, Joseph Bates, Brian Roley, Margaret Luongo, Cathy Wagner, and Michele Navakas each shared their expertise, care, and craft.

I'm also thankful to my (and my parents') high school English teacher Dixie Marshall, who has read and given me feedback on each of my books. Her dedication to her former students is as legendary as her red pen.

Michael O'Donnell and Dave Ring have been my farm gurus for years and have helped guide me over chats, meals, texts, and calls: "Are my missing chickens off living a Disney adventure and they'll come back, or did something take them?" I'd be lost without their input and expertise.

Books that attempt to bring a global perspective are time-consuming and costly to research. Thankfully, Karla Olson and the team at Patagonia Books provided contacts and resources that made this project possible. Even more, Karla has had an enduring vision of what this book could be and has consistently had the patience, guidance, and leadership to get it across the finish line. I'm forever thankful that she put me in the good hands of Sharon AvRutick, who dove into the manuscript when it was double the length of the book you are holding now (!) and, like some kind of book whisperer, helped me make sense of the mass of stories I had compiled. Jane Sievert pored over countless photos to select ones that best served the reader, and Christina Speed pulled them together in this beautiful design. Overall, the Patagonia ecosystem continually introduced me to passionate people who believe a better world is possible, including but not limited to Bernardo Salce and Stephanie Ridge. JL Stermer, my literary agent at Next Level Lit, was always there when I needed a kind word of encouragement.

During the five years it took to live and write this book, the world experienced a global pandemic, I started and completed grad school, and I began working at Ball State University, first in the English Department and then the Honors College. And, if that wasn't enough change and challenge, after a risky surgery cutting into my spinal cord, I was diagnosed with and began receiving treatment for neurosarcoidosis, a rare autoimmune disease that just may be triggered by exposure to chemicals. I could not have made it through any of this without my friends and family, who were not only there for me but also for my wife, kids, and farm animals. They did everything from running the kids around, organizing meal trains, and wrangling rogue sheep to mowing our yard. This group includes Liz Vermeulen, J.R. Jamison, the Truexes, the Moormans, the Wassons, the Taylors, and the Clawsons. Really, there are too many to list here, but you know who you are. Thanks for loving my family.

Aunt Julie and Uncle Steve Amspaugh were always there with hay for the sheep, mashed potatoes and noodles for me, farm tours for the kids, and farm stories and shenanigans from yesteryear.

Getting to spend time with my parents, Ken and Lynne, was the best thing about slowing down during my illness. The deep, perennial small-town roots they and my in-laws, Jim and Gloria Saintignon, have grown regenerated us during the most challenging circumstances. My kids are so lucky to have such active and loving grandparents. Jim passed away in the spring of 2024, but not before setting an example for all of us of the work, joy, and responsibility involved in being a neighbor, friend, father, and spouse.

Lastly, writing a book that required traveling around the country and the world meant I was away from home a lot. Thanks to my amazingly patient and supportive wife, Annie, and our kids, Harper and Griffin, for providing such a wonderful home base. Home is always my favorite destination.

# Index

**A**

Abbott, Christopher, 353–54
açai, 159, 160–61, 167, 168
Acres USA, 293
A-Frame Farm, 96–105
agriculture. *See* Black farmers; industrial agriculture; organic farming; perennial agriculture; permaculture; regenerative agriculture; small farms
Allen, Howard, 263, 264–75, 311, 342
Alliance for a Green Revolution in Africa (AGRA), 135
Aloha 'Aina Poi Company, 218
amaranth, 138
Amazon 4.0 initiative, 151, 159, 160, 163, 164, 170, 176
Amazon River Basin, 147–83
AMABELA, 168–72
Apuoyo wives, 136–38
Arches National Park, 142
Arhuaco people, 2, 12, 45–73, 103, 175, 178, 219, 223, 255, 256, 304, 307, 314, 329–30
Ashkum, Illinois, 350–52
Attenborough, David, 288

**B**

bald eagles, 237, 316
Baldwin, Daryl, 323–24, 338
Baldwin, Jerrid, 325
Bayer, 188
beavers, 63, 251, 308, 333–34, 336–37
beef industry, 245, 248–49. *See also* cattle
bees, 311
Berry, Wendell, 331
Bickford, Paul, 276–87, 290
Bill and Melinda Gates Foundation, 135–36

Black farmers, 267–73
Bluffton, Georgia, 229, 231, 236, 248, 250, 251, 257–61
Bly, Robert, 98
Bolden-Newsome, Chris, 272
Bolsonaro, Jair, 155–56
Brazil nuts, 158
Brosnan, Pierce, 198
Brown, Gabe, 61, 62, 235, 293
Butz, Earl, 86

**C**

cacao, 149–53, 155–59, 161, 168, 177, 181, 183
CAFOs (Concentrated Animal Feeding Operations), 81, 148, 299, 305, 316
Cargill, 135, 166, 183, 245
Carlisle, Liz, 268
Carrboro, North Carolina, 264
Carver, George Washington, 268–70
cassava, 138–39
cattle, 109–22, 125–27, 229, 231–37, 241–46, 248, 250. *See also* beef industry
CEFA (Experimental Active Forest Centre), 181–82
Center for Agricultural Resilience, 240, 259
Chapel Hill, North Carolina, 2
chestnuts, 24, 29–31
chickens, 294, 297–300, 305, 311
Chock, Mason, 211–12
chocolate. *See* cacao
Chumash people, 15
Chun, Malia, 185–88, 190–92, 209, 210, 212–13, 307, 309
cilantro, 295
Cleveland, Scott, 233, 242, 243

PREVIOUS SPREAD: Francisco Vio (left) of Huerto Cuatro Estaciones explains to Kelsey that if he can farm regeneratively in the rugged environment of Chilean Patagonia, it can be done anywhere.
CLIFF RITCHEY

climate change, 9–10, 42, 72, 125, 137, 148, 353–54
coca, 57–58
coffee, 45, 47, 304
Coleman, Eliot, 267, 342
corn, 75, 81–83, 86, 90, 92, 95, 97, 100, 132, 134, 186–89, 194, 198, 203, 318–19
Corry, Stephen, 129
cotton, 52
COVID pandemic, 39, 118, 183, 243, 245, 269, 321
Cumberland Covered Bridge, 316
Cumberland Plateau, 345
cupuaçu, 159, 164, 166–67, 170, 172, 177, 181

**D**

dairy farms, 276–77, 349–50
Dawson, Rosario, 13
DeHaan, Lee, 85–92, 94–96
DeHaan, Robb, 87, 89
Delaware people, 314, 320, 322
De Mendes, César, 149–53, 155–57, 161–64, 166, 170
De Mendes Chocolate da Amazonia, 149, 162–63
desertification, 115–16, 120, 127
direct air capture (DAC) machines, 148
diversity, 137, 263, 267, 332
Dow Chemical, 188
DuPont-Pioneer, 187, 188, 191, 205, 208, 221

**E**

Earthrise Farm, 102
Enonkishu Conservancy, 113–17, 121, 129
Experimental Active Forest Centre (CEFA), 181–82

**F**

Fairfield, Iowa, 349–50
Faithfull Farms, 2, 263, 264–75
Fernholz, Carmen, 96–105, 255, 268, 354

fertilizers, synthetic, 23, 33, 61, 95, 115, 253, 268, 328
food sovereignty, 192, 294
Ford, Henry, 179
Fordlandia, 179
Ford Motor Company, 179
Forever Green Initiative, 89
Fortier, Jean-Martin, 267, 342
Four Seasons Farm (Huerto Cuatro Estaciones), 342–44, 363, 369
Fuhrman, CMarie, 329
fungicides, 75

**G**

Gaston, Indiana, 354–56
Gates, Bill, 135, 260
Gates, Melinda, 135
General Carrera, Lake, 7, 344
GMOs (genetically modified organisms), 23–24, 135, 187, 196, 198, 205
Goodman, Jim, 250
Grain Place Foods, 356–57
Grandin, Greg, 179
grapes, 265–66
Green Revolution 2.0, 131
Greenville, Treaty of, 319

**H**

Half Moon Bay, California, 346–49
HAPA (Hawai'i Alliance for Progressive Action), 193
Harrelson, Woody, 13
Harris, Will, 229–61, 304, 331, 336, 345, 346
Harrison, William Henry, 319–20
Haslett-Marroquin, Reginaldo (Regi), 291–305, 311, 330, 331
Health and Happiness Project, 165, 174, 180, 182
herbicides, 23, 78, 292
Holland, Fern, 197–200, 204, 208, 209, 212
Hooser, Gary, 192–93, 196–98, 200–202, 204–5, 207, 211–13

Howard, Albert, 268
Huerto Cuatro Estaciones (Four Seasons Farm), 342–44, 363, 369

## I

Indigenous people. *See also* individual peoples
    attacks on, 155–56
    knowledge of, 45–47, 72–73, 183, 302, 319
industrial agriculture
    climate change and, 9
    economics and, 79–81, 84–85, 92, 238–39
    food waste and, 82
    fossil fuels and, 33, 92
    future of, 84–85
    land and, 84–87
    negative effects of, 8–10, 14, 78, 83–84, 92, 143–44, 240–41, 244–45, 328–29
    rejection of, 12, 33–34, 137, 185, 268, 332, 356
    seeds and, 75–76, 78, 84, 165
Iniki, Hurricane, 220
insecticides, 23
Itzkan, Seth, 125

## J

Jackson, Wes, 89–94, 105
Janie's Mill, 350–52
JBS, 245
Jeffries, Tyler, 210

## K

Kagawa, Ross, 211
Kahn, Sol, 196–97, 204, 207–8, 210
Kajiado County, Kenya, 118
kalo, 213–18, 220–22, 224–25
Kamehameha, King, 202, 203, 210
Kaua'i, Hawai'i, 5, 12, 185–225
Kernza®, 85, 88, 93–96, 102, 104, 353, 354

Kisumu, Kenya, 131
Konza Prairie, 94
Kumano I Ke Ala Farm, 213–23

## L

Lakota, 257, 347
The Land Institute, 88–94, 104, 354
Lane, Hurricane, 220
Lincoln, Abraham, 14
lions, 109–11, 124
Little Sequatchie River, 345
Lockridge, Ross F., 320, 323
Luo people, 143

## M

Maasai, 12, 109–10, 113–15, 118–21, 124–30, 145, 223, 310, 314, 329
Maasai Center for Regenerative Pastoralism, 121
Madison, Minnesota, 96, 101
Maharishi University, 349
Mainsprings, 38–39
Makua, Kaina, 215–25
Mani, Gilbert Tatanka, 347
Mara River, 112, 114
Mara Training Centre, 121
Marion, Indiana, 317
Markegard, Doniga and Erik, 346–48
Markegard Family Grass-Fed, 346–49
Marquette, Nebraska, 356–57
Matthews, Chris, 354
Matthews Feed and Grain, 316
Mauck, Jason, 354–56
May, Maria Jesus, 343
Maya, 292, 302
Meadowlark Organics, 276–91
Mellencamp, John, 87
Miami people (Myaamia), 308, 314, 318–20, 322–26, 328, 338–39
Michael, Hurricane, 256
Minneapolis, Minnesota, 353–54

# Index

Minnesota, University of, 89
Mississinewa, Battle of, 320, 322, 325
Mississinewa Reservoir, 317, 321
Mississinewa River, 307–9, 313–25, 333, 336, 338–39
Mississippi River, 339
Mollison, Bill, 26
Montgomery, David, 9
Mori, Josh, 191–92, 210–11
Muncie, Indiana, 6
Myaamia (Miami people), 308, 314, 318–20, 322–26, 328, 338–39
Myaamia Center, 324, 325

## N

NAACP, 269
Nabusimake, 56
National Beef Packing Company, 245
National Organic Standards Board, 350
Native Hawaiians, 185–86, 188, 202–3, 223, 314
nature
  cycles of, 11–12, 229, 252–53, 255, 259, 269, 336
  relationship with, 14, 51, 53–58, 62, 65, 68–73, 318–19, 332
Netto, Eugenio Scannavino, 174, 175, 177, 180–83
New Forest Farm, 19–43
noble savage, concept of, 223
Nobre, Ismael, 151, 158, 160, 171–72, 175–77, 180–81
Northfield, Minnesota, 291, 293, 297

## O

O'Donnell, Michael, 12, 331
Ohio River, 328, 339
Organic Consumers Association, 345
organic farming
  certification and standards for, 99, 278, 357
  history of, 99
  time and, 103
  transitioning to, 98
Organic Growers and Buyers Association, 99
origin stories, 62, 325
Otieno, Celestine, 131–45

## P

Patagonia, Chile, 2, 363
Patagonia National Park, 342
Patel, Raj, 250
Penniman, Leah, 269, 272
perennial agriculture, 87–96
Perennial Pantry, 353–54
permaculture, 26, 28
pesticides, 23, 78, 186–88, 191, 196–97, 203–13, 277, 328–29
Pioneer. *See* DuPont-Pioneer
Polihale State Park, 196
Pompermaier, Davide, 164–66, 170
Powers, Richard, 30, 39
Pueblo Bello, Colombia, 48, 51
Puerto Guadal, Chile, 7, 342–44

## R

Radiance Dairy, 349–50
rainforest
  destruction of, 148–49, 160, 168, 174, 179–80
  preservation of, 149, 151, 159–60, 183
Real Organic Project, 350
regenerative agriculture
  beliefs behind, 13–14, 20, 332, 336–37
  cycles of nature and, 11–12, 229, 252–53, 255, 259
  definitions of, 11–12, 252–53, 331–32
  diversity and, 263, 267
  economics and, 356
  giving nature of, 225
  history of, 13, 268
  hope and, 10, 14–15, 332–33

Indigenous knowledge and, 45–47, 72–73, 302, 319
long-term thinking and, 105
misuse of term, 91–92, 331
as permanent agriculture, 28
popularity of, 13, 330–31
standards and, 249, 303–4, 331
time and, 103
transitioning to, 239–40, 250
Regenerative Agriculture Alliance, 300–301
relay cropping, 355
reNature, 149
Ritte, Walter, 191, 208
Robinson family, 202
Rodale, 25, 94, 97, 99, 105, 249, 267–68, 350
Rodale, J.I. and Robert, 267
Rogan, Joe, 249
rubber trees, 179

**S**

Salina, Kansas, 85, 86
sand harvesting, 140–45
Santarém, Brazil, 164–67, 171, 177
Savory, Allan, 115–16, 121, 125, 127, 246, 268
Savory Institute, 121, 125
Sequatchie, Tennessee, 344–46
Sequatchie Cove Farm, 344–46
Seven Pillars, 324–26, 338
sheep, 310, 312, 313
Shepard, Mark, 19–43, 51, 72, 78, 90, 158, 235, 255, 280, 309, 346
Shipman, Pat, 111
Shiva, Vandana, 293
Sierra Nevada de Santa Marta, Colombia, 45–47
small farms
  challenges faced by, 263–64
  profitability and, 132, 134, 267

yields of, 81
Smith, Anthony, 291
snakes, 61–65, 72, 235, 237, 257–58, 292
soil
  importance of, 9–10
  live nature of, 332
  loss of, 10, 142
Soil4Climate, 125
Soler, Javier, 342
Sosoo, Effua, 271–73
Soul Fire Farm, 272
South African Faith Communities' Environment Institute (SAFCEI), 135
soybeans, 75, 81, 86, 90, 95, 97, 165–66, 169
Sprowt Labs, 353–54
Steyer, Tom, 13
STUN (Strategic Total Utter Neglect) method, 19, 33
Survival International, 129
Syngenta, 186, 187, 188, 191, 205

**T**

Tapajós River, 173, 179
taro. *See* kalo
Thicke, Susan and Francis, 349–50
Tiampati, Dalmas, 118–33, 139–45, 222
toads, 46
Tocantins River, 151
Tree-Range Farms, 291–305, 311
Tupinambá, Raquel, 173–78, 180
Tupinambá people, 12, 173–82, 223, 314, 328
Tyson, 243, 245

**U**

UNESCO (United Nations Educational, Scientific and Cultural Organization), 46–47
United Fruit, 302
USDA (US Department of Agriculture), 9, 14, 205, 269–71, 278, 333

# Index

**V**

Van Tassel, David, 94
Vetter, David and Don, 356–57
Vio, Francisco, 342, 363
Viola, Wisconsin, 19, 28, 280

**W**

Wabash River, 325, 339
Waimea River, 214, 225
weeds, definition of, 23
Wepking, Halee and John, 278–91, 346
wheat, 86, 94, 96, 137, 284, 355
White Oak Pastures, 229–43, 249–50, 256–61
Whole Foods, 229, 230, 249–50
Wilken, Harold and Janie, 350–52
Winfrey, Oprah, 349
Woods, Tarquin and Lippa, 113–15, 121

**Y**

Yanomami, 155, 156
Ye'kwana, 155, 156
Young, Neil, 347

NEXT SPREAD: Volunteers roll a giant dibber wheel through the rich, soft soil at Huerto Cuatro Estaciones in preparation for planting. Regenerating Earth and ourselves requires feet in the soil and our hearts and hands at work. CLIFF RITCHEY